CLIMATE CHANGE AND ITS CAUSES, EFFECTS AND PREDICTION

GLACIERS: FORMATION, CLIMATE CHANGE AND THEIR EFFECTS

CLIMATE CHANGE AND ITS CAUSES, EFFECTS AND PREDICTION

Additional books in this series can be found on Nova's website under the Series tab.

Additional e-books in this series can be found on Nova's website under the e-book tab.

CLIMATE CHANGE AND ITS CAUSES, EFFECTS AND PREDICTION

GLACIERS: FORMATION, CLIMATE CHANGE AND THEIR EFFECTS

NICHOLAS DOYLE
EDITOR

nova publishers
New York

Additional color graphics may be available in the e-book version of this book.

Library of Congress Cataloging-in-Publication Data

Names: Doyle, Nicholas, 1978- editor.
Title: Glaciers : formation, climate change and their effects / editor, Nicholas Doyle.
Description: Hauppauge, New York : Nova Science Publishers, Inc., [2016] | Series: Climate change and its causes, effects and prediction | Includes index.
Identifiers: LCCN 2016008524 (print) | LCCN 2016021356 (ebook) | ISBN 9781634849418 (hardcover) | ISBN 9781634849722 ()
Subjects: LCSH: Glaciers. | Glaciers--Climatic factors. | Climatic changes.
Classification: LCC GB2403.2 .G54 2016 (print) | LCC GB2403.2 (ebook) | DDC 551.31/2--dc23
LC record available at https://lccn.loc.gov/2016008524

Published by Nova Science Publishers, Inc. † New York

CONTENTS

PREFACE

This book provides research on the geological formation of glaciers, as well as the effects of climate change to glaciers. The first chapter reviews recent and old applications of first principles physics to the modelling of glaciers and ice sheets. Chapter Two focuses on the physical-geographical factors of the development and evolution of the Debeli Namet glacier in Montenegro. Chapter Three studies high altitude temperature changes in the Tropical Andes over the last 15,000 years. Chapter Four summarizes the current knowledge of the biodiversity and ecology of microbial communities from glaciers. Chapter Five presents the results of investigations of mountain glaciers within the southeastern part of Russian Altai (SE Altai). The last chapter reports a continuation of direct field investigations, aiming to reveal whether supraglacial debris effect is increasing or decreasing in mass balance and glacier evolution with the course of time.

In: Glaciers
Editor: Nicholas Doyle

ISBN: 978-1-63484-941-8

Chapter 1

Glacier Physics from First Principles

Valerio Faraoni*
Physics Department and Environmental Science Programme,
Bishop's University, Sherbrooke, Québec, Canada

Abstract

Theoretical results in glaciology are often obtained (and taught) with rather sophisticated analyses. Sometimes a first principles approach produces nearly the same result in a simple and economical way. We review new and old results, some of which affect estimates of ice volume changes relevant for sea level and climate change. The subjects considered are the densification of snow, the maximum thickness of glaciers and ice sheets, their local thickness, and crevasse depth. Naturally, the discussion differentiates between alpine glaciers on a slope and polar ice sheets.

PACS: *92.40.V-, 92.40.Vq, *92.40.vv, 92.70.Mn

Keywords: longitudinal glacier profiles, ice volume, firn densification, crevasses

1. Introduction

A first principles approach allows for innovative ideas to be developed and, naturally, is usually the clearest approach when introducing people to an entirely new field. In theoretical glaciology, a first principles approach was more

*E-mail address: vfaraoni@ubishops.ca

common in the early days (1950s-1960s). Nowadays the power of modern computers allows for easier and efficient modelling of specific glaciers or ice sheets with a level of detail that was unthinkable in those early days. Does it mean that a first principles approach is now useless? Not really, for one needs to know many details in order to obtain results with numerical simulations and simple models are useful anyway to test elaborate codes and in statistical analyses of large numbers of glaciers [1]. What is more, the beautiful and powerful modelling of specific systems does not provide *general* knowledge expressed in the form of general laws. One may end up with a detailed model of a specific glacier but no new knowledge applicable to any other glacier in the world. Glaciology has evolved more as a data-driven science than a theoretical one, yet any science needs a theoretical framework and general understanding where to accomodate models and data, otherwise it becomes a maze of hyperspecialized subfields which do not talk to each other, sophisticated techniques with restricted applicability, and an overwhelming mass of data and specific models. Without taking away from the specialization needed to achieve results in any branch of science, it is useful to keep in mind a statement by Paterson about the development of glaciology: "... a mere handful of mathematical physicists, who may seldom set foot on a glacier, have contributed far more to the understanding of the subject than have a hundred measurers of ablation stakes or recorders of advances and retreats of glacier terminal" [2].

Even glaciology textbooks largely ignore the simplification brought about by a first principles view when it would be appropriate. This is regrettable because earth science students would benefit from viewing simple ideas before diving into lots of technical details, especially when the latter can be avoided. Indeed, dimensional analysis, zero-dimensional models, and order of magnitude estimates are common in engineering when an exact treatment becomes too cumbersome, or in order to obtain insight before attempting detailed modelling [3, 4, 5, 6].

Here we review recent and old applications of first principles physics to the modelling of glaciers and ice sheets. The subjects considered include the densification of firn, the maximum thickness of glaciers, their local thickness, and crevasse depth.

2. Densification of Firn

One of the most basic processes in glaciology is the metamorphosis of snow into ice, which causes the density of firn in the upper layers of a glacier to depend on the depth while, once the ice density is reached, glacier ice is universally treated as incompressible [2, 7, 8, 9]. Changes in the volume of the giant ice sheets of Antarctica and Greenland determine sea level variations and can be estimated with satellite radar and laser altimeters [10, 11]. Now, volume changes must be converted into mass changes for a reliable estimate of sea level change, which requires knowledge of the density in the upper layers of glaciers and of its changes [12, 13, 14, 15]. Modern interferometric radar techniques and neutron probes allow fairly accurate reconstruction of the density profile in the top layers of ice sheets [16, 17], but the use of average densities is rather tricky and does not translate straightforwardly into mass changes [18].

While the physical processes of snow settlement, rounding of the ice crystals, recrystallization and sintering are very complicated and each of them requires detailed modelling, a macroscopic phenomenological approach (which does not even begin to make justice to the complication of microscopic models) is nevertheless interesting. So, let us step back and look at the densification of firn with the little physical intuition allowed by this complicated phenomenon. Consider a glacier sitting on a horizontal plane and let z be the depth measured from the surface vertically downward. Let $\rho(z)$ be the mass density of firn. In conditions of constant temperature and snow accumulation, Sorge's law assumes that ρ is independent of time [19]. Let us restrict ourselves to dry snow, as the presence of liquid water (due to melting and percolation) and refreezing substantially complicate the picture, making the transformation of snow into ice much faster [2, 7]. Clearly, $\rho(z)$ is minimum at the surface $z = 0$, where it assumes the value ρ_S, and it is bounded from above by the density of ice ρ_i. It is natural to require that $\rho(z)$ is a monotonically increasing function of the depth z, with $d\rho/dz > 0$, and that $\rho(z)$ is at least piecewise continuous with its first derivative. Further, one requires $\rho(z)$ to asymptote to its upper bound ρ_i either at a finite depth z_0, or as (formally) $z \to +\infty$, provided that this approach is sufficiently fast (as, e.g., in exponential solutions of most diffusion models widely used in environmental physics [3]). To summarize, we require that, in the interval $0 \leq z < +\infty$,

- $\rho_s \leq \rho \leq \rho_i$, $\lim_{z\to+\infty} \rho(z) = \rho_i^-$, $\rho(0) = \rho_s$, and

- $\frac{d\rho}{dz} > 0$ where this derivative exists,

which imply that

- $\frac{d^2\rho}{dz^2} < 0$ (where it exists) and $\frac{d\rho}{dz}, \ \frac{d^2\rho}{dz^2} \rightarrow 0$ as $z \rightarrow +\infty$.

Further, the gradient of $\rho(z)$ will be larger for loose snow near the surface which can be easily compacted and will decrease with increasing depth as the firn becomes progressively denser and denser. At depths z such that ρ is approximately that of ice, $d\rho/dz$ will essentially go to zero. We can express this physical intuition by writing $d\rho/dz$ as a function of the "density deficit" $\rho_i - \rho(z)$:

$$\frac{d\rho}{dz} = f\left(\rho_i - \rho(z)\right) . \tag{1}$$

Next, one considers the range of values of $\rho_i - \rho$ of interest. New snow has a density $\rho \sim 50-70\,\mathrm{kg}{\cdot}\mathrm{m}^{-3}$, while settled snow has $\rho \sim 100-200\,\mathrm{kg}{\cdot}\mathrm{m}^{-3}$. The density of dry snow at the surface of polar glaciers (at which we can think that the densification process really begins), is reported as $\rho \sim 330\,\mathrm{kg}\cdot\mathrm{m}^{-3}$, while glacier ice attains densities $\rho \sim 830 - 917\,\mathrm{kg}\cdot\mathrm{m}^{-3}$ [2]. The range of densities involved is not large, spanning at most a a little more than a factor of three, contrary to many other geophysical processes which involve several orders of magnitude in the range of the quantities involved. Because of this very restricted range of densities, it is reasonable to linearize the function $f\left(\rho_i - \rho\right)$ in eq. (1), obtaining the ordinary differential equation

$$\frac{d\rho}{dz} = c\left[\rho_i - \rho(z)\right] , \tag{2}$$

where c is a positive constant with the dimensions of the inverse of a length. All solutions of eq. (2) with values in the $\rho < \rho_i$ range automatically satisfy the requirement $d\rho/dz > 0$.

Eq. (2) is solved trivially: the general solution of the homogeneous equation is $\rho_{\mathrm{hom}}(z) = \rho_0\,\mathrm{e}^{-cz}$ (with ρ_0 an integration constant) while a particular solution of the inhomogeneous equation is $\rho = \mathrm{const.} = \rho_i$. The general solution of eq. (2) is therefore

$$\rho(z) = \rho_0\,\mathrm{e}^{-cz} + \rho_i \tag{3}$$

and clearly satisfies the boundary condition $d\rho/dz \rightarrow 0^+$ as $z \rightarrow +\infty$ (and $d^2\rho/dz^2 \rightarrow 0$ as well). By imposing the other boundary condition $\rho(0) = \rho_s$, one determines the integration constant as $\rho_0 = \rho_s - \rho_i$. The required density profile is then

$$\rho(z) = \rho_i - (\rho_i - \rho_s)\, \mathrm{e}^{-cz}\,, \tag{4}$$

where the constant $c = z_0^{-1}$ provides the scale z_0 over which the firn density changes appreciably (Figure 1). Eq. (4) is exactly the empirical relation reported

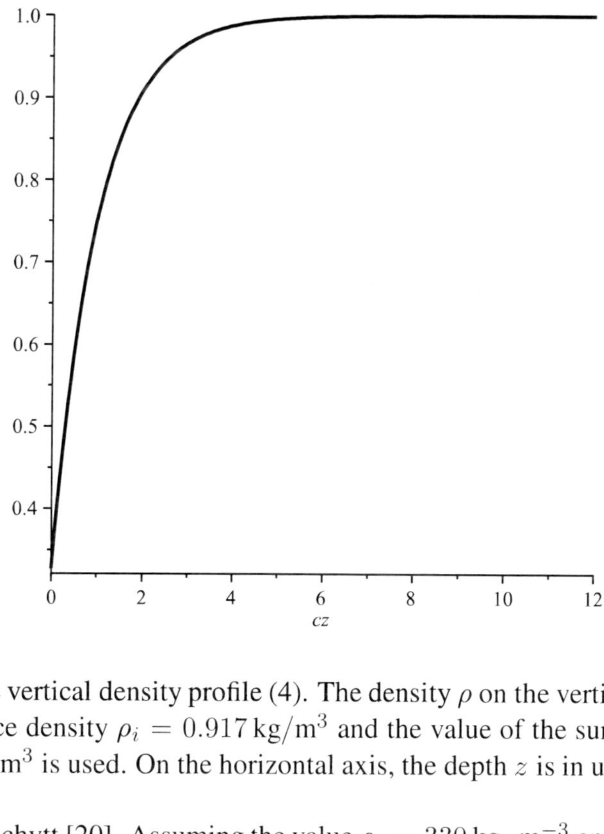

Figure 1. The vertical density profile (4). The density ρ on the vertical axis is in units of the ice density $\rho_i = 0.917\,\mathrm{kg/m^3}$ and the value of the surface density $\rho_s = 300\,\mathrm{kg/m^3}$ is used. On the horizontal axis, the depth z is in units c^{-1}.

long ago by Schytt [20]. Assuming the value $\rho_s = 330\,\mathrm{kg \cdot m^{-3}}$ appropriate for dry snow regions [2], it is $\rho(z_0) = 0.7645\rho_i$, $\rho(2z_0) = 0.9134\rho_i$, and the value $0.99\rho_i$ of $\rho(z)$ is reached at $z = 4.16\, z_0$, hence one can assume that the density of ice is reached at depths $3\, z_0 - 4\, z_0$.[1]

The value of the constant c depends on the stage of densification. For densities up to $\rho_1 \sim 550\,\mathrm{kg/m^3}$ the densification process is more rapid, corre-

[1] By denoting with z_i the *measured* depth of the firn-ice transition, the usual rule of thumb to estimate the value of the constant c at a given location is $c \simeq 1.9\, z_i^{-1}$ [2, 7].

sponding to grain settling and packing by sliding [21, 22, 23], then densification proceeds more slowly with a different value of c until the pores close at $\rho_2 \sim 830\,\text{kg/m}^3$ and air bubbles form. The last phase corresponds to compression of these air bubbles and to yet another value of the constant c [24, 25]. However, the exponential description is less appropriate in the last phase of diagenesis [25]. We stress again that the model described here applies to dry snow: the presence of liquid water due to melting and percolation, and refreezing complicate the picture.

Let $v = dz/dt$ be the vertical velocity of a small volume of firn with position $z(t)$. Mass conservation for a fluid of density ρ and velocity v in a stream tube of cross-sectional area A gives

$$\rho v A = \text{const.} = \rho_s v_s A_s \tag{5}$$

(this is an integrated version of the continuity equation $\partial\rho/\partial t + \vec{\nabla}\cdot(\rho\vec{v}) = 0$). Since there is no lateral motion, one can consider a vertical parallelepiped of constant cross-sectional area $A = A_s$ as the stream tube and

$$\rho(z)v(z) = \text{const.} = \rho_s v_s\,, \tag{6}$$

where the constant quantity $\rho_s v_s$ is the accumulation rate (here in convenient units of $\text{kg}/(\text{m}^2\cdot\text{yr})$). Since $dt = dz/v(z)$ and $v(z) = \rho_s v_s/\rho(z)$, the time taken for a volume of snow deposited at the surface to arrive to the depth z_i of the firn-ice transition is[2]

$$\begin{aligned} t_i &\equiv t(z_i) = \int_s^{z_i} dt = \int_s^{z_i} \frac{\rho(z)}{\rho_s v_s}\,dz = \frac{1}{\rho_s v_s}\left[\rho_i z + \frac{\rho_i-\rho_s}{c}\left(\mathrm{e}^{-cz_i}-1\right)\right]_0^{z_i} \\ &= \frac{\rho_i}{\rho_s v_s}\left[z_i + \frac{1}{c}\left(1-\frac{\rho_s}{\rho_i}\right)\left(\mathrm{e}^{-cz_i}-1\right)\right]. \end{aligned} \tag{7}$$

The age of the ice at the depth z_i of the firn-ice transition can be deduced by counting annual layers of firn near the top of the glacier, like annual growth rings in a tree. By taking $\rho_i = 917\,\text{kg/m}^3$ and $\rho_s = 300\,\text{kg/m}^3$ (a value appropriate for dry, cold snow), it is

$$t_i = \frac{917\,\text{kg/m}^3}{\rho_s v_s}\left[z_i + \frac{0.6728}{c}\left(\mathrm{e}^{-cz_i}-1\right)\right]. \tag{8}$$

[2]There is no need for numerical integration of the z-ρ curve, as stated in Paterson's textbook [2] since the elementary integration above gives an exact result.

Let us apply this formula to the table of measured values of accumulation rates, depths z_i of the firn-ice transition, densification times t_i, and the values of the constant c obtained by least-square fitting of the data, which are reported in Tables 2.2 and 2.3 of Paterson's textbook [2]. We can then compare the calculated age t_i with the measured value. This comparison is not entirely trivial because some assumptions underlie the little calculation leading to eq. (8). However, to the best of our knowledge, this comparison is not presented in the literature.

2.1. Site A (Greenland)

For the Site A location in Greenland (a dry snow area), the experimental data are [2]

$$\rho_s v_s = 265 \, \frac{\text{kg}}{\text{m}^2 \cdot \text{yr}} \,, \tag{9}$$

$$z_i = 75 - 80 \, \text{m} \,, \tag{10}$$

$$t_i = 185 \, \text{yr} \,, \tag{11}$$

$$c = 2.35 \cdot 10^{-2} \, \text{m}^{-1} \,. \tag{12}$$

Using the middle range value $z_i = 77.5$ m of the transition depth, eq. (8) yields $t_i = 185$ yr, in agreement with the measured value.

2.2. Site 2 (Greenland)

Site 2 is another dry snow location, for which the measured values are [2]

$$\rho_s v_s = 410 \, \frac{\text{kg}}{\text{m}^2 \cdot \text{yr}} \,, \tag{13}$$

$$z_i = 66 \, \text{m} \,, \tag{14}$$

$$t_i = 120 \, \text{yr} \,, \tag{15}$$

$$c = 2.46 \cdot 10^{-2} \, \text{m}^{-1} \,. \tag{16}$$

For these data eq. (8) gives $t_i = 98.5$ yr, with a significant disagreement with the measured value.

2.3. Dome C (Antarctica)

Dome C in Antarctica is another dry snow zone, with experimental data [2]

$$\rho_s v_s = 36 \, \frac{\mathrm{kg}}{\mathrm{m}^2 \cdot \mathrm{yr}} , \tag{17}$$

$$z_i = 100 \, \mathrm{m} , \tag{18}$$

$$t_i = 1700 \, \mathrm{yr} , \tag{19}$$

$$c = 1.65 \cdot 10^{-2} \, \mathrm{m}^{-1} . \tag{20}$$

Eq. (8) yields $t_i = 1708$ yr, in good agreement with the measured value.

2.4. Maudheim (Antarctica)

The well studied site of Maudheim in Antarctica is not a dry snow area but corresponds to the upper percolation zone, with measured values [2]

$$\rho_s v_s = 370 \, \frac{\mathrm{kg}}{\mathrm{m}^2 \cdot \mathrm{yr}} , \tag{21}$$

$$z_i = 67 \, \mathrm{m} , \tag{22}$$

$$t_i = 125 \, \mathrm{yr} , \tag{23}$$

$$c = 2.58 \cdot 10^{-2} \, \mathrm{m}^{-1} . \tag{24}$$

Here eq. (8) yields $t_i = 113$ yr.

2.5. Byrd (Antarctica)

At Byrd, another upper percolation zone, the experimental data are [2]

$$\rho_s v_s = 140 \, \frac{\text{kg}}{\text{m}^2 \cdot \text{yr}} , \tag{25}$$

$$z_i = 64 \, \text{m} , \tag{26}$$

$$t_i = 280 \, \text{yr} , \tag{27}$$

$$c = 2.75 \cdot 10^{-2} \, \text{m}^{-1} , \tag{28}$$

and eq. (8) provides the age $t_i = 286$ yr, in good agreement with the measured value.

2.6. Little America (Antarctica)

At Little America (upper percolation zone) the data are [2]

$$\rho_s v_s = 220 \, \frac{\text{kg}}{\text{m}^2 \cdot \text{yr}} , \tag{29}$$

$$z_i = 51 \, \text{m} , \tag{30}$$

$$t_i = 150 \, \text{yr} , \tag{31}$$

$$c = 3.14 \cdot 10^{-2} \, \text{m}^{-1} , \tag{32}$$

which, using eq. (8), yield $t_i = 141$ yr.

As we can see, the exponential model (4) which can be derived phenomenologically with little effort, agrees reasonably well with the measures in the field.

3. Maximum Thickness of a Glacier or Ice Sheet

A problem that seems to have partially escaped attention in the development of glaciology and that surprisingly does not appear in textbooks is that of the maximum possible thickness of a continental glacier or polar ice sheet, while the corresponding problem for alpine glaciers is standard material, although the emphasis in the literature is never on the maximum possible thickness. It is well known that the measured thickness of glacier ice ranges from tens of meters in valley glaciers to almost 5 km in Antarctic ice sheets [2, 7]. Given sufficient accumulation of snow, a glacier will advance and eventually reach a state of equilibrium with a typical average thickness. What physics sets the upper limit of $\sim$ 5 km to the maximum thickness of ice sheets? This bound should be predicted by basic physics, but this question has been answered only recently in an article in a pedagogical physics journal [26] (the upper limit of tens or hundreds of meters for valley glaciers is instead discussed in textbooks [2, 7, 8]). This fact is rather surprising because an almost identical question appears in most geography books and popular earth science documentaries and publications about *mountains* instead of glaciers. How high can a mountain be? Every geographer or earth scientist has a ready answer: if a mountain grows too high, it will start melting at its base and flow. This argument will be applied here to ice sheets, but first it is essential to note that a distinction needs to be made about valley glaciers and polar ice sheets. In a valley glacier, gravity will "break" the ice long before it can grow so thick that it starts melting at its base. In the following we make this distinction, reviewing and expanding the arguments of [26].

3.1. Alpine Glaciers

For completeness, and for contrasting with the next subsection, here we review some standard material. Consider a glacier resting on a slope, such as an alpine glacier: the upper limit to the ice thickness is set by gravity. Intuitively, one can imagine isolating an ideal block of ice in a glacier on a uniform slope and calculating the component of the force of gravity along the slope which tends to make it fall: this pull must be compensated by internal stresses to hold it in place (fig 2). However, the magnitude of these internal stresses is not unlimited but is set by the microscopic electrostatic bonds between water molecules in the solid state, i.e., eventually, by quantum mechanics. If the gravitational force on a surface (a side of the ideal block) inside the ice exceeds the tensile stress, the ice block cannot stay in place and will begin to move. This is a very crude way of reasoning but it is correct as a first approximation. Giant ice avalanches in which a significant fraction (if not most) of a ramp glacier breaks off are

Figure 2. Internal stresses keep these ice "blocks" up on the Miage glacier on Mont Blanc, France.

famous in the history of glaciology and mountain environments (we are not discussing the perennial fall of seracs in icefalls here). An example is the giant 1895 ice avalanche of the Altels glacier (Valais, Switzerland) estimated at $\sim 4 \cdot 10^6$ m^3, which is to date the largest ice avalanche recorded in the European Alps. Another tragic example is that of the Allalin glacier (Valais, Switzerland), which collapsed on August 30, 1965 releasing $\sim 2 \cdot 10^6$ m^3 of ice and causing 88 victims among the workers at the Mattmark hydroelectric plant dam 400 m below (this event and the dynamics of the glacier are modelled in [27]). This was not an isolated event either, as the glacier had broken off before.

Let us consider, therefore, a block of ice of thickness h, basal area A, with nearly parallel sides sitting on a uniform slope inclined at an angle α with the horizontal and sheared by the action of gravity (Figure 3). As a first approximation, one neglects basal sliding[3] and assumes that the forward motion of the

[3]This approximation would be incorrect, for example, for the 1965 Allalin event, in which basal lubrication due to a hot summer is believed to be crucial in the modelling [27].

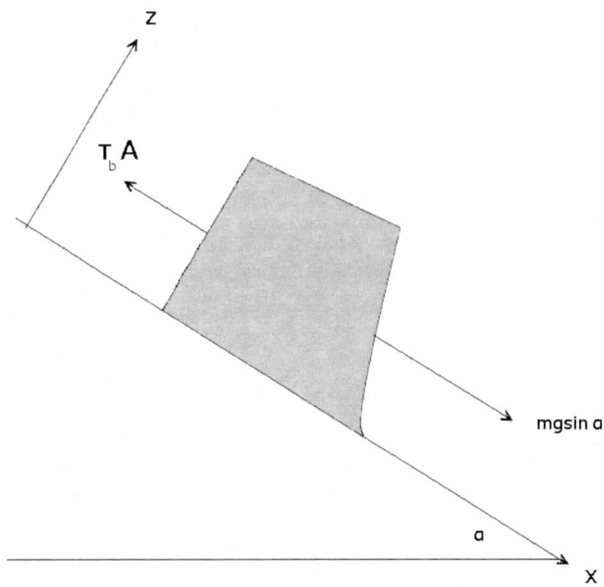

Figure 3. A block of ice with basal area A on a slope, subject to the component of the force of gravity parallel to the slope $mg\sin\alpha$ and to the tensile stress $\tau_b A$.

glacier is due only to the internal deformation of ice. Let τ_b be the basal (shear) stress of the ice: the force holding the ice block in place has magnitude $A\tau_b$ and must balance the component of the block's weight which is parallel to the slope and pointing downward, $mg\sin\alpha$. Here g is the acceleration of gravity and $m = \rho h A$ is the mass of the block, while $\rho_i = 917\,\text{kg/m}^3$ is the ice density. Equilibrium between those two forces gives the block thickness as[4]

$$h = \frac{\tau_b}{\rho_i g \sin\alpha} \, . \tag{33}$$

A typical value is $\tau_b \sim 10^5$ Pa, which yields [2, 7]

$$h = \frac{11\,\text{m}}{\sin\alpha} \, , \tag{34}$$

[4]Glaciology textbooks presenting this formula [2, 7, 8] are usually more interested in obtaining a value for the basal stress τ_b, which is difficult to measure, than in the thickness h of the ice.

a reasonable value in agreement with field observations. This value is corrected for valley glaciers by introducing an empirical dimensionless shape factor $f \sim 0.5 - 0.9$ which takes into account the support given by the valley walls to the ice [2, 7],

$$h = \frac{\tau_b}{f \rho_i g \sin \alpha} = \frac{11 \, \text{m}}{f \sin \alpha} \, . \tag{35}$$

Thus far, this is standard textbook material: it can also be shown [2, 7] that the angle α at the base of the glacier can be exchanged with the angle that the surface of the ice makes with the horizontal (which is much easier, although still not trivial, to measure), in the approximation that these angles are small.

3.2. Ice Sheets

In the limit $\alpha \to 0$ of a horizontal glacier, the maximum ice thickness (33) diverges, meaning that gravity does not set limits on this quantity on a flat surface. However, there is definitely an upper limit ~ 5 km on Earth, hence some other physics intervenes to set such a bound. The argument about the maximum height of a mountain (approximately 10 km on Earth) can be applied, *mutatis mutandis*, to this problem, therefore it is useful to revisit it. Although standard lore in geography and earth science, the argument is rarely attributed or is its source mentioned. The argument is due to Victor Weisskopf, one of the fathers of modern quantum physics, who discussed the problem in an article in *Science* in 1975 [28]. Weisskopf was interested in the strength of the moleular bonds and saw the macroscopic height of a mountain as a giant confirmation of microscopic physics determined by molecular quantities. It is fair to speculate that Weisskopf must have been attached to his simple, clear argument because he reiterated it in the pedagogical physics journal *American Journal of Physics* eleven years later [29] and replied [30] to an objection raised in the same forum [31]. The argument for a mountain goes like this: approximate the mountain as a parallelepiped of S_iO_2. When a layer of this material added to the top causes plastic flow at the bottom, the mountain has reached its maximum height. The gravitational potential energy lost by the top layer lowered by the height h supplies a fraction η of the energy needed to melt a layer of this material with the same thickness at the base of the mountain. Let the parallelepiped have basal area A, height h, volume $V = Ah$, and mass $M = \rho V$ (where ρ is the density of the material), while the layer has thickness δ and mass $m = \rho A \delta$. The gravitational potential energy lost by this layer is $mgh = \rho A g h \delta$, which is equated to ηL_f, where L_f is the latent heat of fusion of S_iO_2. Weisskopf obtains [28, 29, 26]

$$h = \frac{\eta\, L_f}{g} = 14\,\text{km} \tag{36}$$

using $L_f = 1.65 \cdot 10^6\,\text{J/kg}$ and $\eta = 1/3$ (using granite or basalt instead of S_iO_2 gives the same order of magnitude [26]).

The argument can be applied to glacier ice by taking the view (explicitly expressed in many geology textbooks) that ice can be seen as a rock near its melting point. By repeating the argument using the numerical values $\eta \sim 1/3$ and $L_f = 3.34 \cdot 10^5\,\text{J/kg}$, one obtains $h \simeq 11$ km [26]. This is the correct order of magnitude for polar ice sheets.

4. Local Thickness of an Ice Sheet

Here we follow again an argument of Ref. [26]. Imagine slicing an ice sheet from its maximum height (H) at the summit to its terminus along an x-axis running through its base. The thickness profile $h(x)$ thus obtained, or *local* thickness, is a longitudinal glacier profile [2, 7, 8, 9], with $0 \leq h(x) \leq H$. Consider, as a simple model, a slab of ice on a bed idealized as a plane sloped to form an angle β with the horizontal. Let this slab of ice have unit width in the transversal (y) dimension. The x-axis along this incline points downward and a z-axis, perpendicular to it, points upward. A classic one-dimensional problem of glaciology consists of modelling the local thickness $h(x)$ of the ice. Different assumptions about the physics of glacier ice lead to different longitudinal profiles, includinge Nye's parabolic profile [32, 33], Vialov's profile [34], and Bödvardsson's and Bueler's profiles [35, 36]. Two of these are discussed separately in the two subsections below.

The profile of a glacier is relevant for climate change considerations because, given the area of a glacier (deduced, e.g., from aerial surveys), the profile gives the ice volume and is used to estimate volume changes related to climate change or rising seal levels.

4.1. Perfectly Plastic Ice

A first possible assumption is that of perfectly plastic ice. This is clearly an idealization but, because of the semplicity it brings, it was widely used in early approaches to theoretical glaciology [32, 33, 2, 7] and its results are still widely used in modern analyses.

In equilibrium, the component of the weight of the ice slab of mass m, which is $mg \sin \beta$, must be balanced by the basal shear stress. Under the usual approximation replacing α with β for small slopes [2, 7], one has

$$\sin \beta \simeq \sin \alpha \simeq \tan \alpha = -dh/dx \,. \tag{37}$$

The condition that the resultant of the two forces on the slab vanishes becomes

$$\tau_b A \simeq -\rho g h A \frac{dh}{dx} \,. \tag{38}$$

By imposing the initial condition that the thickness h vanishes at the glacier terminus $x = L$, one obtains the initial value problem

$$\frac{d}{dx}\left(\frac{h^2}{2}\right) = -\frac{\tau_b}{\rho g} = \text{const.}\,, \tag{39}$$

$$h(L) = 0 \,, \tag{40}$$

which has the trivial solution [26]

$$h(x) = H\sqrt{1 - \frac{x}{L}} \,, \tag{41}$$

where

$$H = \sqrt{\frac{\tau_b}{\rho g}}\sqrt{L} \tag{42}$$

is the maximum thickness. The exact treatment by Nye [32, 33] is not very different from what shown here. The only difference between Nye's exact parabolic profile and the approximate one described by eqs. (41) and (42) consists of a factor $\sqrt{2}$ multiplying the maximum thickness, which becomes $H = \sqrt{\frac{2\tau_b}{\rho g}}\sqrt{L}$ in the exact solution [32, 33].

4.2. Ice Satisfying Glen Law

Again, a simplification of the problem of the local thickness of a glacier is introduced by assuming that the ony cause of motion is the flow of ice and neglecting basal sliding and deformation of subglacial till. Let $c(x)$ be the volume of precipitation (assumed here, for simplicity, to be instantly converted into ice) per unit of normal area and per unit time, i.e., the flux density of ice volume. The volume $c(x)$ of ice crossing the unit of area normal to the z-axis per unit time at the location x (in $\frac{\mathrm{m}^3}{\mathrm{m}^2\cdot\mathrm{s}}$) is, essentially, the negative[5] of the component u_z of the three-dimensional velocity of the ice $\vec{u} = (u_x, 0, u_y)$, so $c(x) = -u_z(x)$. Assuming steady state equilibrium, the ice loss is equal to the gain and mass conservation is expressed by the continuity equation

$$\frac{\partial \rho}{\partial t} + \vec{\nabla} \cdot (\rho \vec{u}) = 0\,, \tag{43}$$

where ρ denotes the density of ice. Assuming, as usual in glacier flow modelling, the ice to be incompressible, eq. (43) reduces to $\vec{\nabla} \cdot \vec{u} = 0$. Using the order of magnitude approximation

$$\frac{\partial u_x}{\partial x} \simeq \frac{\bar{u}}{x} \tag{44}$$

(where $\bar{u}$ is the average velocity of the ice flow in the x-direction u_x) and

$$\frac{\partial u_z}{\partial z} \simeq \frac{-c(x)}{h(x)}\,, \tag{45}$$

the reduced continuity equation $\frac{\partial u_x}{\partial x} + \frac{\partial u_z}{\partial z} = 0$ becomes

$$\frac{\bar{u}}{x} - \frac{c}{h} \simeq 0\,. \tag{46}$$

The resulting relation $cx = h\bar{u}$ between vertical and horizontal motions is well known in glacier modelling, where it is derived, instead, using a vertically integrated form of the continuity equation (43) ([2, 7, 8, 9], see the discussion in Ref. [26]).

In the previous section we assumed perfectly plastic ice. Let us consider now the more realistic description of isotropic ice by Glen's law [37]. Let σ_{ij} be the stress tensor inside the ice and let $s_{ij} = \sigma_{ij} - \frac{\mathrm{Tr}(\hat{\sigma})}{3}\,\delta_{ij}$ (in Cartesian coordinates) be the trace-free part of σ_{ij}, i.e., the deviatoric stress. The effective shear stress, i.e., the invariant

$$s_{\text{eff}} = \left[s_{xx}^2 + s_{yy}^2 + s_{zz}^2 - \left(s_{xx}s_{yy} + s_{yy}s_{zz} + s_{xx}s_{zz}\right)\right]^{1/2}\,, \tag{47}$$

[5] Here the z-axis points upward while the vertical velocity of the ice, which is downward, is negative.

is now assumed to be related to the strain rate $\dot{\epsilon}_{ij}$ by Glen's law [37]

$$\dot{\epsilon}_{ij} = \mathcal{A} s_{\text{eff}}^{n-1} s_{ij} \,, \tag{48}$$

where $\mathcal{A}$ is a (temperature-dependent) constant and the value $n = 3$ is appropriate for ice flow [37, 2, 7, 8]. We use our first principles approach and following again [26], the strain rate is

$$\dot{\epsilon}_{ij} = \frac{1}{2}\left(\frac{\partial \dot{\xi}^i}{\partial x^j} + \frac{\partial \dot{\xi}^j}{\partial x^i}\right) = \frac{1}{2}\left(\frac{\partial u^i}{\partial x^j} + \frac{\partial u^j}{\partial x^i}\right) \,, \tag{49}$$

where $\vec{\xi}(\vec{x})$ is the displacement of the ice at the position $\vec{x}$ due to the applied stress and an overdot denotes differentiation with respect to time. In particular, for the component $\dot{\epsilon}_{xz}$ one has

$$\dot{\epsilon}_{xz} \simeq \frac{1}{2}\frac{\partial u_x}{\partial z} = \mathcal{A}\sigma_{xz}^n \,. \tag{50}$$

Using now, in the same approximation as before,

$$\frac{\partial u_x}{\partial z} \simeq \frac{\bar{u}}{h} \tag{51}$$

the relation $cx \simeq h\bar{u}$ gives us [26]

$$cx \simeq h\bar{u} = h^2 \frac{\partial u_x}{\partial z} = 2h^2 \mathcal{A}\sigma_{xz}^n \simeq 2\mathcal{A}h^2 \tau_b \simeq 2\mathcal{A}h^2 \left(-\rho g h \frac{dh}{dx}\right)^n \,. \tag{52}$$

We end up with the non-linear ordinary differential equation (expressing the highly non-linear behaviour of ice described by the Glen law (48))

$$cx = 2\mathcal{A}h^2 \left(-\rho g h \frac{dh}{dx}\right)^n \tag{53}$$

with $n = 3$. This equation, obtained from first principles (plus the empirical Glen law as an essential ingredient) [26] is very close to the Vialov equation [34] obtained with a rigorous derivation

$$cx = \frac{2\mathcal{A}}{n+2} h^2 \left(-\rho g h \frac{dh}{dx}\right)^n \,, \tag{54}$$

the difference being just the factor $1/(n+2)$ on the right hand side of eq. (54). For the value $n \simeq 3$ appropriate for ice flow, this factor is equal to 0.2 (however, this factor appears to the power $1/n$ in the solution $h(x)$, the difference becoming a multiplicative factor 0.58).

5. Crevasse Depth

To conclude our list of examples of first principles physics applied to glaciology, let us discuss briefly crevasse depth. Crevasses form in a glacier when at least one of the principal stresses (pointing normally to the crevasse walls) is a tensile stress and exceeds the tensile strength of ice [2, 7]. The theoretical study of crevasses and the problem of the maximum depth of a crevasse constitute an application of the vast field of fracture mechanics (*e.g.*, [38]), see the brief discussion of Ref. [8] for the essential references.

A simple order of magnitude estimate of the depth of a crevasse, however, does not require the more complicated formalism and results of fracture mechanics and is due to Nye [39]. Consider a crevasse which is open down to a depth h (Figure 4).

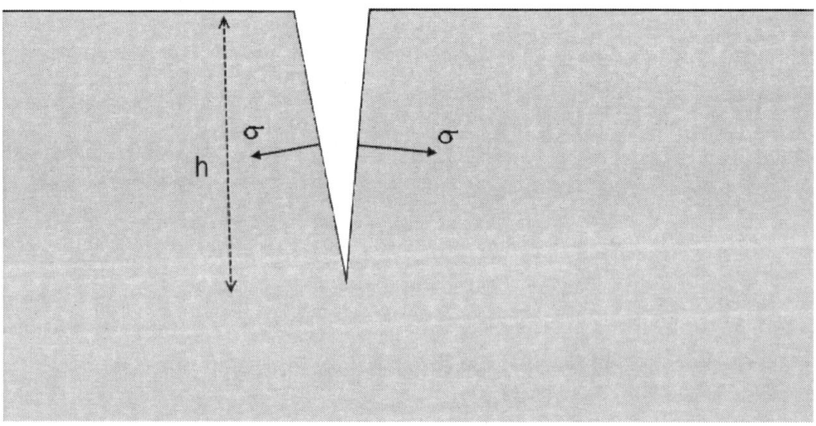

Figure 4. A crevasse of depth h with tensile principal stress σ applied to each wall.

Assume that equal principal stresses σ contribute to opening the crevasse on each side, therefore the widening action is due to a total stress 2σ. The cryostatic pressure instead tends to close the crevasse and the maximum crevasse depth will be reached when the cryostatic pressure equals twice the principal stress [39],

$$\rho_i g h \simeq 2\sigma \,, \tag{55}$$

which gives

$$h \simeq \frac{2\sigma}{\rho_i g} \simeq 22\,\mathrm{m} \tag{56}$$

using the values $\sigma \sim 10^5$ Pa for the yield stress of ice, $\rho_i = 917\,\mathrm{kg/m^3}$ for the ice density, and $g = 9.8\,\mathrm{m/s^2}$ for the acceleration of gravity. Reports of crevasse depths in temperate glaciers rarely exceed $h \sim 30$ m, which means that Nye's simple order of magnitude estimate is basically correct. However, the yield stress increases as the temperature decreases, hence crevasses in cold polar glaciers can be deeper than in temperate glaciers and, indeed, there are reports of larger depths in cold polar regions [2, 7, 8].

6. Conclusion

We have illustrated with a few examples how first principles physics reproduces quickly, with little effort, and without the need of much theoretical baggage, certain new and classic results of glaciology. Of course, first principles, orders of magnitude estimates, and dimensional analysis are no substitute for a full treatment of complicated systems, but they should always be performed before the investment of a detailed and sophisticated analysis is made. Furthermore, a first principles analysis constitutes a sanity check when details become cumbersome in "real" treatments, it provides simple models to test long computer codes and for use in statistics, and it is an excellent way to introduce a new field to beginners. This is particularly true when fields of research such as climate change and glaciology, which have numerous connections with other areas and involve nonlinear interacting systems, are studied simultaneously. It is regrettable that first principles considerations often are not seen as valuable and are not reported in the literature or are dismissed altogether. An example discussed here is the problem of the maximum thickness of an ice sheet, which seems to have gone unnoticed until recently, in spite of the fact that the analogous problem for mountains is reported so often even in the non-technical literature. The fact that certain physical arguments are not commonly used by glaciologists does not legitimate suspicion or belief that they are wrong. Indeed, there is not only beauty, but also power in simplicity and first principles analysis on the side of complete treatment should be encouraged whenever possible.

References

[1] Jouvet, G., Rappaz, J., Bueler, E. and Blatter, H. 2011. "Existence and stability of steady-state solutions of the shallow-ice-sheet equation by an energy minimization approach". *J. Glaciol.* 57: 345-354.

[2] Paterson, W.S.B. 1994, *The Physics of Glaciers*, 3rd edition. Oxford: Butterworth-Heinemann.

[3] Boeker, E. and van Grondelle, R. 2011, *Environmental Physics*. Chicester: J. Wiley & Sons.

[4] Harte, J. 1998, *Consider a Spherical Cow: A Course in Environmental Problem Solving*. Sausalito, CA: University Science Books.

[5] Harte, J. 2001, *Consider a Cylindrical Cow: More Adventures in Environmental Problem Solving*. Sausalito, CA: University Science Books.

[6] Johnson, J.N. 1974. "Kinematic waves and group velocity: applications to natural and man-made environment". *Am. J. Phys.* 42: 681-688.

[7] Cuffey, K.M. and Paterson, W.S.B. 2010, *The Physics of Glaciers*. Amsterdam: Elsevier.

[8] Hooke, R. LeB. 2005. *Principles of Glacier Mechanics*, 2nd edition. Cambdridge, UK: Cambridge University Press.

[9] R. Greve and H. Blatter 2009, *Dynamics of Ice Sheets and Glaciers*. New York: Springer.

[10] Wingham, D.J., Shepherd, A., Muir, A. and Marshall, G.J. 2006. "Mass balance of the Antarctic ice sheet". *Phil. Trans. Roy. Soc. A* 364: 1627-1635.

[11] Pritchard, H.D., Arthern, R.J., Vaughan, D.G., and Edwards, L.A. 2009. "Extensive dynamic thinning of the margins of the Greenland and Antarctic ice sheets", *Nature* 461: 971-975.

[12] Braithwaite, R., Laternser, M., and Pfeffer, W. 1994. "Variations of near-surface firn density in lower accumulation area of the Greenland ice sheet, Pakitsoq, West Greenland". *J. Glaciol.* 40: 477-485.

[13] Arthern, R. and Wingham, D. 1998. "The natural fluctuations of firn densification and their effect on the geodetic determination of ice sheet mass balance". *Clim. Change* 40: 605-624.

[14] Zwally, H. and Li, J. 2002."Seasonal and interannual variation of firn densification and ice-sheet surface elevation at the Greenland summit". *J. Glaciol.* 48: 199-207.

[15] Zwally, H., Giovinetto, M.B., Li, J., Cornejo, H.G., Beckley, M.A., Brenner, A.C., Saba, J.L., and Yi, D. 2005. "Mass changes of the Greenland and Antarctic ice sheets and shelves and contributions to sea-level rise: 1992-2002". *J. Glaciol.* 51: 509-527.

[16] Arthern, R.J., Corr, H.F.J., Gillet-Chaulet, F., Hawley, R.L., and Morris, E.M. 2013. "Inversion of the density-depth profile of polar firn using a stepped-frequency radar", *J. Geophys. Res.: Earth Surface* 118: 1257-1263.

[17] Hawley, R.L., Morris, E.M., and McConnell, J.R. 2008. "Rapid techniques for determining annual accumulation applied at Summit, Greenland", *J. Glaciol.* 54: 838-845.

[18] Li, J. and Zwally, H.J. 2011. "Modeling of firn compaction for estimating ice-sheet mass change from observed ice-sheet elevation change", *Ann. Glaciol.* 52: 1-7.

[19] Bader, H. 1954. "Sorge's law of densification of snow on high polar glaciers", *J. Glaciol.* 2: 319-323.

[20] Schytt, V. 1958."Glaciology II. Snow studies at Maudheim — Snow studies inland — The inner structure of the ice shelf of Maudheim as shown by core drilling", Norwegian-British-Swedish Antarctic expedition 1949-52. Scientific Results vol. 4: A, B, C.

[21] Benson, C.S. 1962. "Stratigraphic studies in the snow and firn of the Greenland ice sheet", US Snow, Ice and Permafrost Research Establishment, Research Report 70.

[22] Anderson, D.L. and Benson, C.S. 1962. "The densification and diagenesis of snow", in *Ice and snow: properties, processes and applications*, Pro-

ceedings, MIT February 12-16, 1962, W.D. Kingery *et al.* eds., pp. 391-411. Cambridge, Mass.: MIT Press.

[23] Alley, R. 1987. "Firn densification by grain-boundary sliding: a first model", *J. Phys. Colloq.* 48: 249-256.

[24] Bader, H. 1965. "Theory of densification of dry, bubbly glacier ice", US Cold Regions Research and Engineering Laboratory Research Report 141.

[25] Herron, M.H. and Langway, C.C. 1980. "Firn densification: an empirical model", *J. Glaciol.* 25: 373-385.

[26] Faraoni, V. and Vokey, M.W. 2015. "The thickness of glaciers". *Eur. J, Phys.* 36: 055031.

[27] Failletaz, J., Funk, M., and Sornette, D. 2012. "Prediction of alpine glacier sliding and instabilities: a new hope", Proceedings, EGU General Assembly Vienna, Austria 22-27 April 2012, p. 7243 [preprint arXiv:1201.1189].

[28] Weisskopf, V.F. 1975. "Of atoms, mountains, and stars: a study in qualitative physics". *Science* 187: 605-612.

[29] Weisskopf, V.F. 1986. "Search for simplicity: mountains, water waves, and leaky ceilings", *Am. J. Phys.* 54: 110.

[30] Weisskopf, V.F. 1986. "Physics of mountains—response". *Am. J. Phys.* 54: 871.

[31] Landis, G.A. 1986. "Physics of mountains", *Am. J. Phys.* 54: 871.

[32] Nye, J.F. 1951. "The flow of glaciers and ice-sheets as a problem in plasticity". *Proc. Roy. Soc. Lon. A* 207: 554-572.

[33] Nye, J.F. 1952. "A method of calculating the thicknesses of the ice-sheets", *Nature* 169: 529-530.

[34] Vialov, S.S. 1958. "Regularization of glacial shields movement and the theory of plastic viscous flow", *IAHS Publ. 47*, Symposium at Chamonix 1958, Physics of the Movement of the Ice, Wallingford, UK: IAHS Press.

[35] Bödvardsson, G. 1955. "On the flow of ice-sheets and glaciers", *Jokull* 5: 1-8.

[36] Bueler, E., Lingle, C.S., Kallen-Brown, J.A., Covey, D.N., and Bowman, L.N. 2005. "Exact solutions and verification of numerical models for isothermal ice sheets". *J. Glaciol.* 51: 291.

[37] Glen, J.W. 1955. "The creep of polycrystalline ice". *Proc. R. Soc. Lon. A* 228: 519-538.

[38] Lawn, B. 1993. *Fracture of Brittle Solids*. Cambridge, UK: Cambridge University Press.

[39] Nye, J.F. 1952. "The mechanics of glacier flow". *J. Glaciol.* 2: 82-93.

In: Glaciers
Editor: Nicholas Doyle
ISBN: 978-1-63484-941-8

Chapter 2

PHYSICAL-GEOGRAPHICAL FACTORS OF ORIGIN AND EVOLUTION OF THE DEBELI NAMET GLACIER (MONTENEGRO)

Predrag Djurović[1,*] and Mirela Djurović[2]
[1]Geography Faculty, University of Belgrade, Belgrade, Serbia
[2]Faculty of Science, University of Novi Sad, Novi Sad, Serbia

ABSTRACT

The Debeli Namet glacier is located on the Durmitor Mountain in Montenegro. It is considered to be the southern most glaciers in Europe and the only modern glacier in overall area of Dinarids. In size is micro, i.e., embryonic glacier.

The significant role in its origin and evolution had its macro, and particularly micro positioning respect to large relief landforms. Pleistocene cirques were recently under karst processes modified and transformed into karst forms (uvalas). Such closed forms considerably supported the accumulation of snow and formation of numerous snow patches as well as the Debeli Namet glacier. Geomorphological factors that supported the development of the Debeli Namet glacier are low

[*]Corresponding author: Predrag Djurović. Studentski trg 3/III, 11000 Belgrade, Serbia. E-mail: geodjura@eunet.rs.

temperatures (cryogenic), huge amounts of snow (nival) and carbonate basement (karst).

In the highest elevations of the Durmitor Mt., as well as at other high mountains in Montenegro dominates periglacial process. On carbonate basement formed recent protalus ramparts, while those formed during the Little Ice Age survived. At non-carbonate basement persisted rock glaciers. According to recent morphology and climate conditions the hypothetical boundary of glacial process has been established to beneath the highest mountain peaks on Durmitor. As glacial process takes part above a periglacial zone, the glacier Debeli Namet represents the out of zone occurrence.

Certain variations in glacier size were noted during instrumental observations. Analyses of two main climate factors (air temperature and precipitation) for the last four decades reflected on certain climate changes that had positive and negative influence on size of the Debeli Namet glacier. The most important changes were in its thickness while the length and the areal extent remained less affected due to its specific position.

Keywords: Debeli Namet, glacier, Dinarids, Montenegro

INTRODUCTION

A number of recent glaciers in the south Europe, i.e., Mediterranean area are small glaciers (Grünewald, 2010). The recent ELA above the altitude of recent glaciers in this part of Europe (Hughes, et al., 2009), is referring them to azonal occurrence. The origin of recent glaciers, as well as the factors that lead to their fluctuation, make them particularly interesting. Within the last 20-30 years these glaciers underwent more or less changes related to their aerial extent and thickness. The reduced thickness caused some glaciers to break off and split into two parts (Ghiacciaio del Calderone, Central Apennines) (D'Orefice et al., 2000; Dramis et al., 2002, D'Alessandro et al., 2003; Pecci, 2006), or to melt completely and to regenerate afterwards (Triglav glacier, Julian Alps) (Gabrovec, et al., 2014).

Recent glaciers in south Europe were studied on Durmitor Mountain (Djurović, 1996; 1999; 2011; 2012; Gachev et al., 2012; 2016; Hughes, 2007; 2008; 2010), Prokletije (Milivojević et al., 2007; Hughes, 2008), Pirin (Grünewald et al., 2008; 2008a; Grünewald et al., 2010; Gachev, 2009; Gachev et al., 2009a), Slovenian Alps (Šifrer, 1976; Tintor, 1993; Gams, 1994; Gabrovec et al., 2014), Italian Apennines (D'Orefice et al., 2000;

Dramis et al., 2002; D'Alessandro et al., 2003; Pecci, 2006), Pyrenees (Grove, 2004), Sierra Nevada, Spain (Gómez et al., 2003), Picos de Europa (González-Trueba, 2004), and Alps Maritime (Federici et al., 2001).

The existence of recent glacier in Dinaric Alps was not known till the mid-20th century (Nicod, 1968). It was finally confirmed during the study of the high mountain karst on Durmitor Mountain since 1987 to 1996 (Djurović, 1996). Intensive researches within the last 30 years included several topics. The contents of heavy metals and radioactive elements in ice were determined in the Debeli Namet glacier (Veselinović et al., 1997), as well as natural and anthropogenic influences as threats to pits in the glacier (Djurović, 1999); radiochemical methods, namely, γ-spectrometry (137Cs isotope) and β-activity measurements (3 H concentration by LSC technique) was applied on ice cores extracted from glacier's ice (Kern, 2006). The relative ages of the youngest moraines built by the Debeli Namet glacier were determined using the lichenometry method (Hughes, 2007). The aim of these studies was also the influence of extreme summer temperatures on the Debeli Namet glacier extent (Hughes, 2008). The given glacier was additionally the scope of researches and monitoring of future behavior of the South Europe glaciers (Grunewald, 2010; Coluccia et al., 2015).

Recognized changes in the extent of the Debeli Namet glacier, which happened within the last 50 years, are of notably less intensity in respect to other glaciers in this area (Coluccia et al., 2015). This is the consequence of different physical-geographic background (geological background and orography), as well as of contemporary climate variances, particularly in air temperature and precipitation regime. One of significant factors is also the way of glacier is feeding (ratio between snowfalls, avalanches and wind-drifted snow).

For better understanding the origin and development of the Debeli Namet glacier including causes and extent of recent changes, and their comparison with another similar micro glaciers in given area, is necessary to introduce physical-geographic factors that contributed to its behavior.

Geographical Position

The Debeli Namet glacier is located on the south-easternmost part of the Dinaric Mountain chain, in central part on Durmitor Mount in the north of Montenegro. Durmitor is the highest mountain in Dinaric system. Bobotov kuk (2523 m a.s.l) is the highest peak, and the lowest point is at Sušičko Lake at about 1150 m a.s.l. Mountain is surrounded by high karst plateaus distributed from 1400 to 1600 m a.s.l. altitude (Figure 1).

Durmitor is bordered on west by the Sušica canyon, the Komarnica canyon on south and by the high Jezerska plateau on south. One third of the mountain's area is up to elevation of 1800 m a.s.l., while the remaining three thirds exceed 1800 m a.s.l. (Djurović, 2011). The distance from the coast of Adriatic Sea is approximately 100 km. Such outlined mountain occupies the space between 18°59´ and 19°07´ east longitude and between 43°05´ and 43°12´ north latitude, covering the area of 96 km^2.

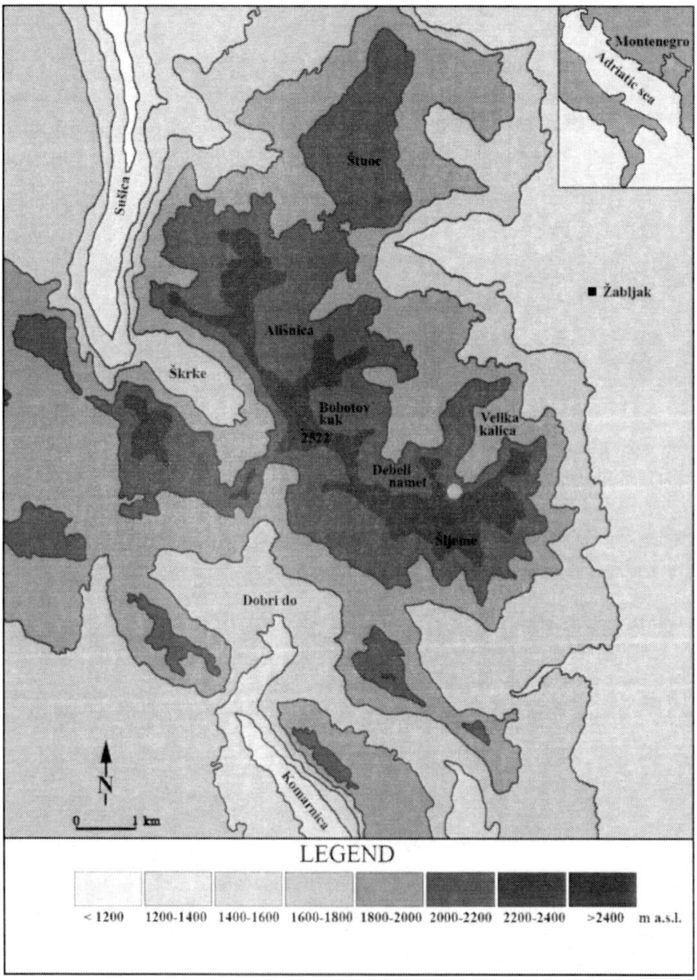

Figure 1. Geographical position of the Debeli Namet glacier.

The whole mountain area belongs to the National park "Durmitor," which is on the UNESCO World Heritage List of natural and cultural sites. In administrative sense is the mountain shared between the Šavnik, Plužine and Žabljak municipalities. The last is the administrative and cultural center of this region and is located at the very east mountain slope.

The Mount Durmitor is deeply incised by Pleistocene cirques and glacier valleys, with remarkably high and commonly vertical ridges between. In the highest altitude of one of such cirques - Velika kalica is the Debeli Namet glacier located.

Terminal moraine that borders glacier on the lowest altitude at 2050 m a.s.l. is synchronously the lowest glacier point. The highest glacier point varies year by year from 2150 to 2180 m a.s.l. This enables glacier to reach 100 to 130 m in height. Its length is from 320 to 350 m.

Geological Characteristics of the Study Area

Durmitor is mostly built of carbonate rocks. Clastic rocks of Mesozoic and Cenozoic age are notably subordinated (Mirković, 1983). Carbonate complexes compose different limestones of Mesozoic and Cretaceous-Paleogene age (Figure 2). Massive to banked limestones of Jurassic age are represented with all three stages and are the most abundant on the Durmitor Mountain, occurring as narrow either continual or dismembered zones. Interlayers of cherts are common. Jurassic limestones build northern, eastern and southeastern parts of the mountain. Cretaceous sediments include massive and banked limestones of pronounced thickness and purity. They are exposed at the highest central mountain parts. The youngest carbonate rocks are Cretaceous-Paleogene sediments, known as the Durmitor flysch. It comprises banked and bedded limestones, marly limestones, carbonate breccias, etc. which outcrop on the southern mountain parts. Clastic rocks are less abundant and occur sporadically. The commonest are Lower Triassic quartz sandstones and sandy limestones. Middle Triassic andesites occur also. All mentioned rock types occupy eastern mountain slope. Quaternary sediments, represented by rare Pleistocene moraine material and recent talus accumulations are insignificantly abundant.

Pleistocene cirque Velika Kalica, in which is the Debeli Namet glacier placed formed within the massive Middle Triassic zoogenic reef limestones with corals and brachiopods. Limestones are few hundred meters thick, very pure and cut by numerous deep cracks and faults (Mirkovic, 1983).

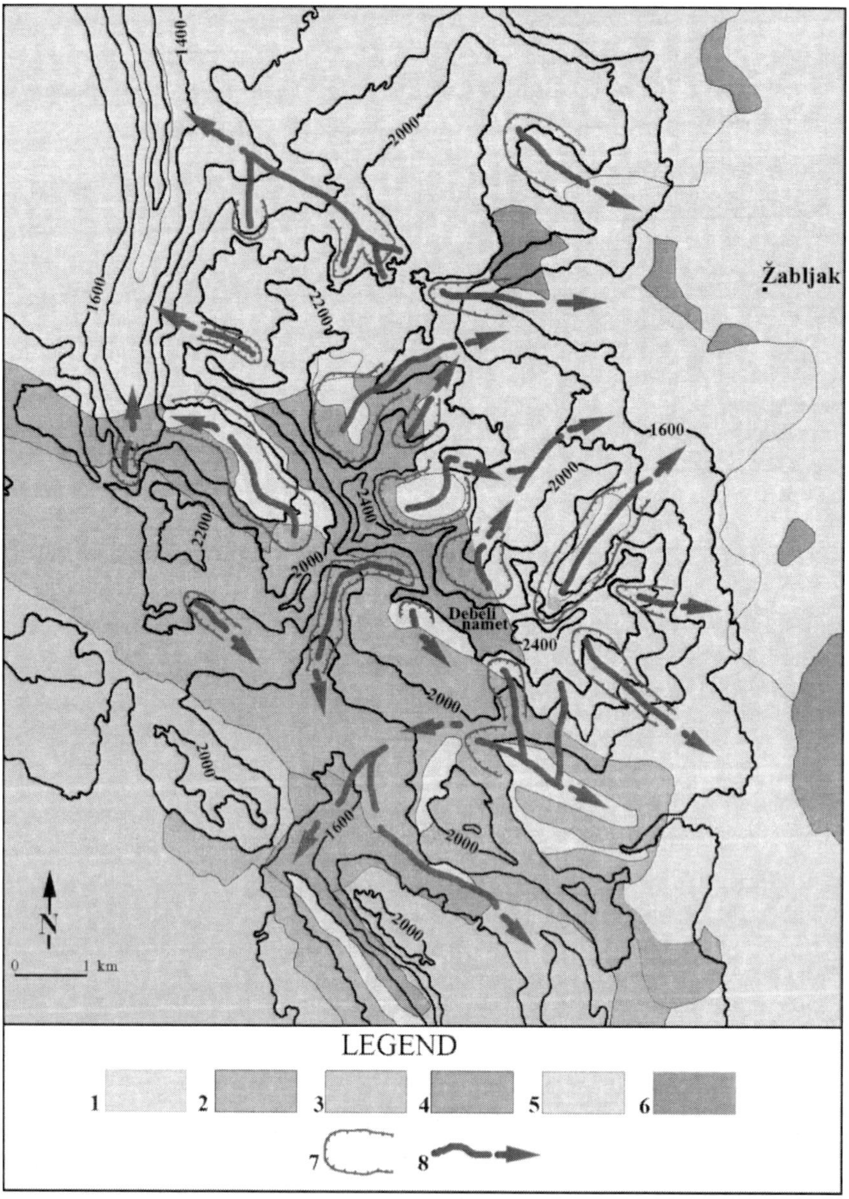

Figure 2. Geological-geomorphological map of Mt. Durmitor: 1 – Quarternary sediments, 2 - Cretaceous-Paleogene carbonate, 3 - Cretaceous limestones, 4 – Jurassic limestones, 5 - Triassic limestones, 6 - Clastic rocks (sandstones, andesites), 7 – Cirque, 8 – Moving direction of Pleistocene glaciers.

GEOMORPHOLOGY OF THE STUDY AREA

The main characteristic of modern relief on Durmitor is its development under the influence of different geomorphological processes that used to alternate frequently during Quaternary. Though, polygenetic relief forms were developed throughout the alternation of different geomorphological processes (Figure 2).

Glacial process took part during glacial phases, while in interglacial periods it has been replaced by karst once including periglacial process, which's more or less influence governed the elevation. Consequently, the modern relief on Durmitor displays properties of karst relief with elements of glacial and periglacial relief.

The most authors that explored glacial relief on this mountain within the last 120 years have reached agreement that it formed during three glacial phases: older, younger and cirque (Cvijić, 1889; 1903; Milojević, 1937; 1951; Marović et al., 1972; Menković et al., 2004; Milivojević, 2007; Djurović, 2002; 2009;). The oldest phase was the strongest, and after it the intensity of glacial process continually weakens. A number of glacial phases was determined mainly according to erosive glacial forms, and just partly by glacial accumulations (Hughes, 2011).

Glacial sediments served as the base for determination of the absolute age of glacial phases. Although the recognition of glacial accumulations was based on lithological criteria and included sediment properties such as clast characteristics (e.g., size, lithology, shape, roundness, presence or absence of striae), clast fabric, matrix composition (e.g., particle size) and sedimentary structures (Hughes, 2011) a problem of subjective approach to recognition and afterwards interpretation, has remained (Djurović, 2012). Secondary carbonates (calcite) cementing the glacial sediments were dated using Uranium-series. According to these studies the strongest glaciation took part before 350 ka, during the Middle Pleistocene. Such statement was estimated in correlation with vast glacial phase elsewhere in Europe that took part during MIS-12 (470-420 ka). The next, less intensive phase corresponds to a period MIS 10-6 (360-130 ka). Ice fields and valley glaciers phase, notably weaker from previous one, took part during MIS 5d-2. Finally, the last glacial phase (cirque and valley glaciers) happened during the Younger Dryas (12.9-11.7 ka) (Hughes, 2011).

According to erosion glacial forms and complex geomorphological analyze of relief the reconstruction of ELA for the three mentioned glacial phases was made (Djurovic, 2009).

During the older glacial phase the ELA was at 1600 m a.s.l. on northern, eastern and southeastern slopes of Durmitor. In southern mountainous part it descended up to a 1400 m a.s.l., whereas on north was approximately at 1500 m a.s.l. The average value for ELA in this glacial phase is at 1540 m a.s.l.

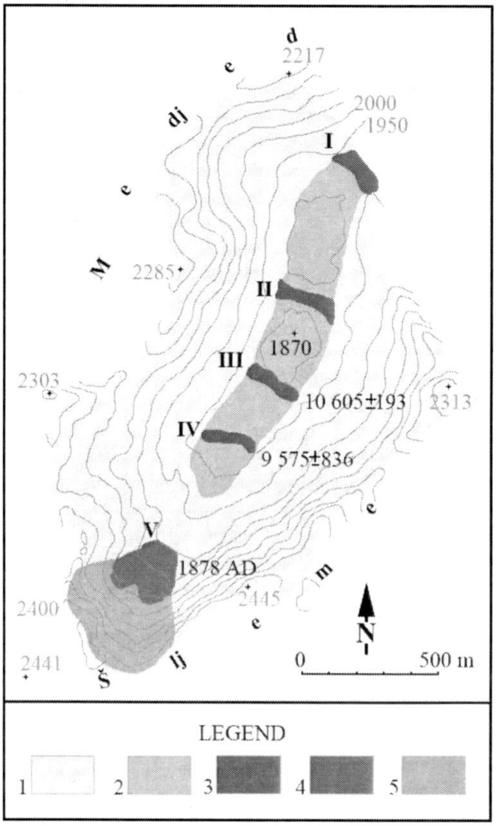

Figure 3. The drainage area for the Velika kalica: 1 - Sides of the Pleistocene cirque, 2 – Bottom of the Pleistocene cirque, 3 – Moraines, 4 – Bottom of the recent cirque of the Debeli Namet glacier, 5 – Sides of the recent cirque of the Debeli Namet glacier.

During the younger glacial phase the average ELA was on north at 1850 m a.s.l., in central part of the mountain at 1900 m a.s.l., and at 1850 m a.s.l. on east. The average value for ELA for the whole mountain in younger glacial phase was at 1880 m a.s.l.

Cirque phase marked the further increase of ELA. It reached 1950 m a.s.l. in northern mountain parts, and 2100 m on east (where the recent glacier Debeli Namet has been developing). On west and southwest is ELA at about

2080 m a.s.l., and at 1950 m a.s.l. on northwest. During this glacial phase the average ELA was at 2020 m a.s.l.

Fluctuations in the intensity of glacial processes led to changes in extent of mountainous area covered with ice. Therefore, in the strongest glacial phase about 54% of this mountain was under the ice; in younger about 36%, while in the last cirque phase only 3.2% of the mountain remained glacier-covered. The Debeli Namet glacier is occupying only 0.025% of the mountain spatial area at present (Djurovic, 2009).

Plenty cirques formed during Pleistocene have being modified in Holocene into uvalas. According to modification degree was distinguished three evolution types of uvalas on Durmitor: glacial, glacial-karst and karst-fluvial. Previous glacial morphology is well-preserved in glacial type. Such uvalas are more or less opened in the direction of past glacier advance. Their sides are exposed to intensive colluvial process enabling a huge amount of talus material to accumulate in the lowest altitudes. Intensive karstification caused fast sinking of surface waters leaving behind a surface lack of flows.

The Debeli Namet glacier is located in the upper part of the Velika Kalica uvala (Figure 3). Its length is 1050 m, width 200 m. Its bottom is at 1870 m a.s. l. and the aerial extent 0.14 km^2. Five moraine ridges occur in it (Djurović, 2011). Four lower moraines derived by successive retreat of glacier during the last glacial phase, while the highest, fifth moraine is modern one. The age of Pleistocene moraines was determined using the U-series method on secondary calcite cement. The third moraine at 1910 m a.s.l. is 9.575 ± 836 years old. For the fourth moraine, placed above (1930 m a.s.l.) was determined age of 10.605 ± 193 (Hughes et al., 2008).

The fifth and the highest moraine is a recent one (2050 m a.s.l.). It consists of limestone fragments ranging in size from few cm to few meters. The color and morphological differentiation of loose material within the recent moraine enabled three ramparts to be distinguished (Djurović, 1996; Hughes, 2007). Two bigger ramparts are in outer part of moraine, and the third one is on internal side (Figure 7). Terminal part of moraine is lack of ridges. All three ridges were dated by lichenometry method due to rate of lichen colonies, precisely of lichen species Aspicilia calcarea agg (Hughes, 2007). Relationship between size and age of lichens on gravestones served for determination the age for the fifth, the highest moraine in Velika Kalica. Average diameter for the five largest lichens at the external ridge is 86.8 mm and 67.5 mm on interior ridge. Linear regression of obtained values for lichens on monuments and gravestones of known ages (Matthews, 2005 (Jomelli et al., 2007) led to their dating. The first ridge in outer part of moraine, as has been estimated, started to be colonized by lichens in 1887, and another one in 1903. As the ridge at interior side of moraine is lack of lichens it was considered not to be older than 11 years (Hughes, 2007).

CLIMATE CHARACTERISTICS OF THE STUDY AREA

Analyses of climate elements involved analyses of air temperature and precipitation as being of the highest influence on the glacier Debeli Namet dynamics. The air temperature was either instrumentally measured where it was possible, or statistically estimated on sites lack of instrumentally equipment.

Meteorological station "Žabljak" is the only station on mountain. It is located on the eastern mountain slope, in the city of Žabljak at 1450 m a.s.l.

The average air temperature in Žabljak for reference period 1961-1990 is 4.7°C (Table 1) and 5.3°C for the whole observation period (1958-2010). The average air temperature in Žabljak tended to decrease till the beginning of 80s in the last century, and afterwards gradually and continually increases. The last two decades displays further increase of the air temperature, but with noteworthy annual fluctuations (Figure 4).

The results obtained throughout instrumental measurements in the meteorological station Žabljak, due to periphery position and small altitude should not be accepted as being valid for the overall mountain area. Therefore, the statistical calculations were applied in order to determine approximate values of the air temperature and precipitation in the highest mountain area.

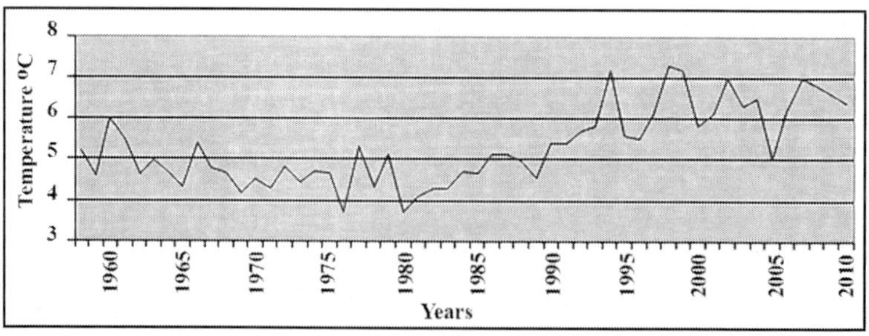

Figure 4. Average annual air temperatures (meteorological station Žabljak).

Table 1. Average monthly and annual air temperatures (1961-1990) (meteorological station Žabljak)

Months	I	II	III	IV	V	VI	VII	VIII	IX	X	I	XII	Annual
Temperatures	- 4.3	- 3.7	- 1.0	3.1	8.5	12.0	14.0	13.7	10.0	5.9	1.3	- 2.1	4.7

Table 2. Vertical temperature gradient on Durmitor

Monthly	I	II	III	IV	V	VI	VII	VIII	IX	X	XI	XII	Annual
TG	- 0.334	- 0.555	- 0.660	- 0.658	- 0.559	- 0.449	- 0.430	- 0.419	- 0.455	- 0.418	0.430	- 0.283	- 0.437
+/- °C	0.100	0.152	0.247	0.127	0.077	0.063	0.058	0.058	0.067	0.078	0.135	0.097	0.090
KF	- 0.985	- 0.741	- 0.487	- 0.789	- 0.878	- 0.902	- 0.910	- 0.909	- 0.986	- 0.876	- 0.775	- 0.854	- 0.856

TG – vertical temperature gradient; +/- °C – analytical error; KF – correlation factor.

Table 3. Estimated mean monthly and mean annual air temperatures at selected altitudes

Alt. m a.s.l.	I	II	III	IV	V	VI	VII	VIII	IX	X	XI	XII	Annual
1550	- 4.6	- 4.3	- 1.7	2.4	7.9	11.6	13.6	13.3	9.5	5.5	0.9	- 2.4	4.3
1650	- 5.0	- 4.8	- 2.3	1.8	7.4	11.1	13.4	12.9	9.1	5.1	0.4	- 2.7	3.8
1750	- 5.3	- 5.4	3.0	1.1	6.8	10.7	12.7	12.4	8.6	4.6	0.0	- 2.9	3.4
1850	5.6	5.9	3.6	0.5	6.3	10.2	12.3	12.0	8.2	4.2	- 0.4	-3.2	3.0
1950	- 6.0	- 6.5	- 4.3	- 0.2	5.7	9.8	11.9	11.6	7.7	3.8	- 0.9	- 3.5	2.5
2050	- 6.3	- 7.0	- 5.0	- 0.8	5.1	9.3	11.4	11.2	7.3	3.4	- 1.3	- 3.8	2.1
2150	- 6.6	- 7.6	- 5.6	- 1.5	4.6	8.9	11.0	10.8	6.8	3.0	- 1.7	- 4.1	1.6
2250	- 7.0	- 8.1	- 6.3	- 2.2	4.0	8.4	10.6	10.3	6.4	2.6	- 2.1	- 4.4	1.2
2350	- 7.3	- 8.7	- 6.9	- 2.8	3.5	8.0	10.1	9.9	5.9	2.1	- 2.6	- 4.6	0.8
2450	- 7.6	- 9.3	- 7.6	- 3.5	2.9	7.5	9.0	9.5	5.5	1.7	- 3.0	- 4.9	0.3
2550	- 8.0	- 9.8	- 8.2	- 4.1	2.4	7.1	9.3	9.1	5.0	1.3	- 3.4	- 5.2	0.1

Determination of vertical temperature gradient on Durmitor (Table 2) included data from meteorological stations Žabljak (1450 m a.s.l), Pljevlja (784 m a.s.l), Nikšić (647 m a.s.l), Bijelo Polje (606 m a.s.l), Kolašin (944 m a.s.l). Linear correlation between altitudes and air temperature beneath 0.6 reflected on insufficient dependence of these two parameters. Data from meteorological station in Nikšić was excluded from further estimations due to its opening and subjection to warm currents from the Adriatic Sea and Skadar Lake. The repeated linear correlation indicated a prominent dependence of these two parameters (0.95). In the first phase were determined vertical temperature gradients for mean monthly temperature and mean annual air temperature between the Žabljak station and stations Pljevlja, Bijelo Polje and Kolašin. After that the average values for mean monthly and mean annual temperature gradient (TG) were calculated. For the obtained gradient value the linear correlation with monthly and annual values, as well as the correlation factor, were determined.

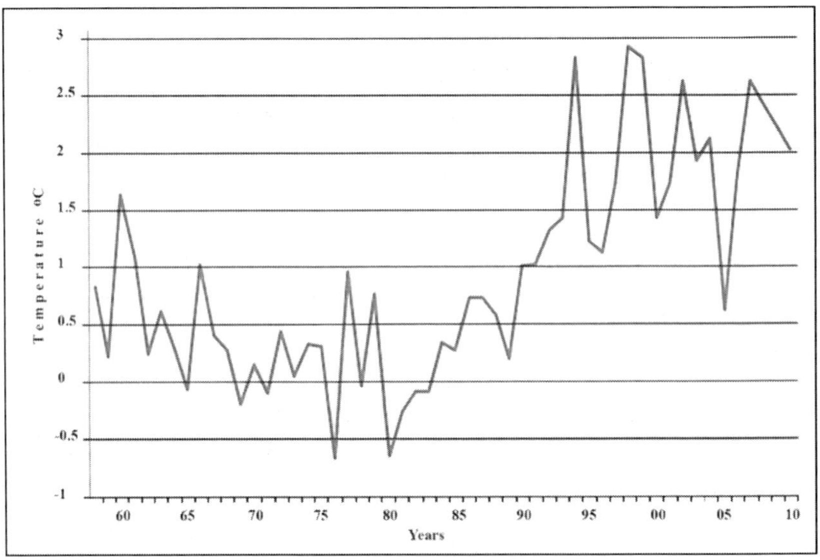

Figure 5. Average annual air temperature at elevation of 2450 m a.s.l.

Average annual temperature of 2.1°C is obtained for the elevation of the lowest point of the glacier Debeli Namet (2050 m a.s.l.) (Table 3). Average monthly temperature within the six months is negative. However, the estimated average annual temperature at 2450 m a.s.l. has been negative since the beginning of 1970s to mid-1980s (Figure 5). The highest altitude of the cirque Velika Kalica, where the accumulation and transport of snow towards the glacier Debeli Namet takes part, may be with negative average annual air temperature during the same period (Figure 6).

Thc air temperature in meteorological station (MS) "Žabljak" was changing throughout decades as follows:

Years	51-60	61-70	71-80	81-90	91-00	01-10
°C	5.1	4.7	4.5	4.7	5.4	6.0

Thereby it can receive much more precipitation than other parts of the mountain. Analyzing the fluctuation in precipitation throughout decades reflected on values listed below:

Years	61-70	71-80	81-90	91-00
mm	1514.2	1564.4	1287.5	1370.1

Analyses of precipitation included data submitted from eight meteorological stations, distributed over the mountain at elevations from 446 to 1280 m a.s.l. along with the data from meteorological station "Žabljak."

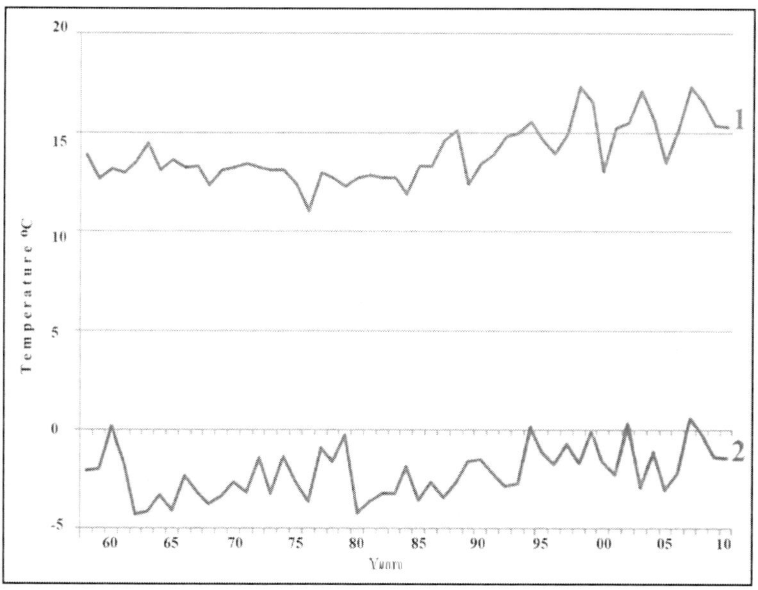

Figure 6. Average summer (1) and winter (2) air temperatures (meteorological station "Žabljak").

Distribution and altitudes of stations enables insight into the precipitation rate and regime along the mountain rim. However, the highest central part of the mountain is lack of any instrumental measurements. In order to obtain adequate values of precipitation rate for the whole mountain area the statistical analyses were applied. Such obtained data should be taken in consideration with reserve as local conditions, relief dissection, extension of canyon valleys and mountain ridges in respect to direction of moisture air masses could significantly affect the rate of precipitation.

Average precipitation in Žabljak is 1450 mm. In any of the analyzed stations the dependence of precipitation from altitude was not noted. Perpendicular position of mountain to a direction of moisture air current motions from Mediterranean is the key factor of precipitation. Consequently, the stations on southwestern slopes record for 60-70% more precipitation then stations on north-eastern mountain side (rain shadow). According to these calculations the central parts of Durmitor (above 2000 m a.s.l.) receive approximately 2600 mm of precipitation annually (Djurović 1996; 1999; 2011; 2012).

The ridge Šljemen, which borders the cirque Velika Kalica where is the glacier Debeli Namet placed, by its southern and eastern margins represents the first and the mostly exposed ridge on southwestern Durmitor slopes. Therefore, it can receive much more amount of precipitation then the rest of the mountain.

Extreme values of certain climate elements could illustrate the complex climate fluctuations within the last ten years. The last decade has been the warmest in Žabljak since the implementation of instrumental measurements. However, in February 28, 2005 was reported the absolutely highest snow depth of 2.30 m in Žabljak. The earliest first snow since the implementation of instrumental measurements (in 1958) swept in July 9, 2006, reaching 13 cm in depth. In forenoon hours in July 23, 2007 in Žabljak flurried the snow whereas above the elevations of 2000 m a.s.l. formed the snow layer. Absolute maximal air temperature in Žabljak is 31.3°C, and was recorded in July 24, 2009. The absolute monthly precipitations were recorded in November (591 mm) and December (415 mm) in 2010 (Republic Hydro-Meteorological Institute of Montenegro-RHZCG).

The glacier Debeli Namet feeds directly from snowfalls but avalanches and wind-drifted snow contribute significantly.

Glaciological Studies

The aim of performed glaciological studies and observations was to define the glacier with its qualitative properties and size, and to highlight fluctuations in size, advance velocity and way of feeding.

According to changes in dipping and to characteristics of ice-snow mass at longitudinal profile of glacier were distinguished three parts: lower (concave), medium (convex) and upper (concave) part (Figure 6).

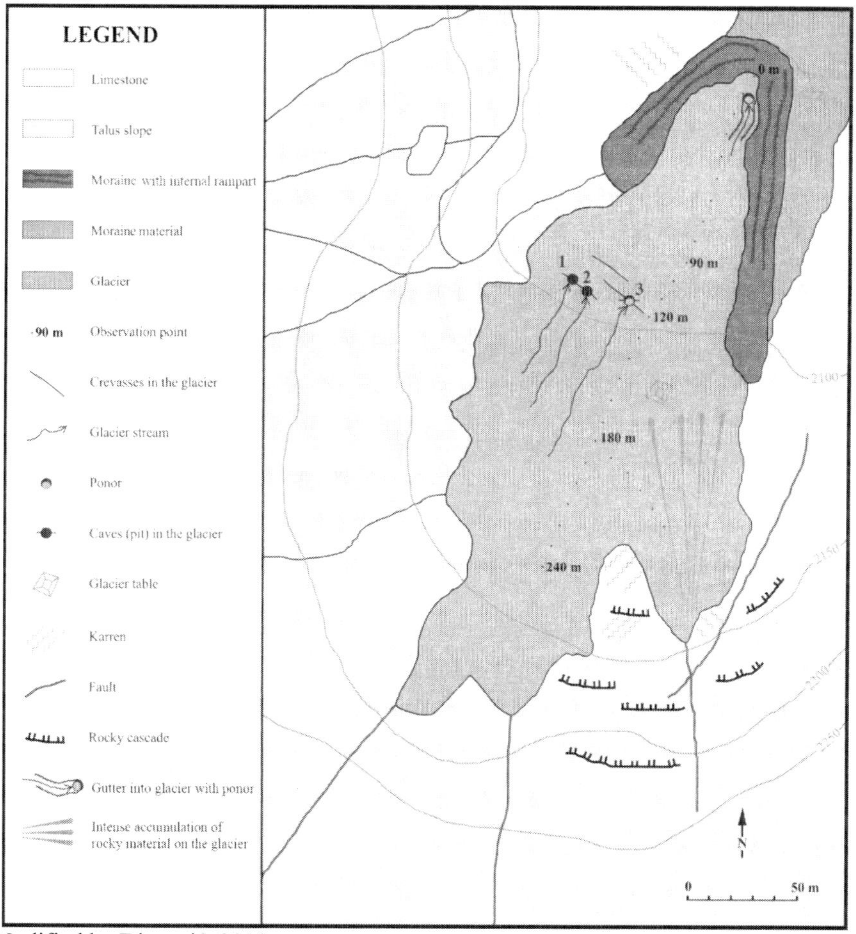

Modified by Djurović, 2012.

Figure 6. The Debeli Namet glacier.

Lower glacier part (glaciers tongue) is at the lowest side bordered by vertical wall of the terminal moraine. The highest point of the terminal moraine is at 2050 m a.s.l. what is the altitude that glacier reaches during winter period (accumulation period). Seasonal melting of snow reduces the thickness of the glacier, i.e., snow and firn of it. Annual variation, about 4 m, in thickness of the glacier is actually the difference in altitudes of the glacier advance (2046 m a.v.l.).

Middle part of the glacier represents the beginning of its tongue and is determined by changes in angle of dip of the glaciers surface (turns from concave into convex). Line of change is about 120 m distant from the terminus of the glacier.

The upper part is feeding area of the glacier, i.e., the area where snow accumulates turning into glacier's ice. Feeding area occupies space from line of change in dip and transition from convex into concave shape at about 120 m from terminus. Such dip the glacier will retain till its end, i.e., till some 320 to 350 m away from terminus (2150-2180 m a.s.l.).

Figure 7. 1. The recent moraine, 2. External and internal ramparts of recent moraine, 3. Cirque of the Debeli Namet glacier, 4. Gutter in the glacier, 5. Layers in the ice mass. (Photographs by P. Djurović)

In the upper, northwestern part of the glacier a three flows evolve from water released by melting of snow during summer. They flow over surface and sink in the middle of the glacier, probably across the oblique cracks inside the glacier (Figure 6). At places where water sinks the farther melting of ice led to creation of three caves (Figure 8). Their depth of 8 m is the thickness of the glacier, also. At cave bottoms occurs debris (subglacial moraine). These caves represent the truly natural sonde as being pulled across the overall glacier mass.

The vertical analyze of the glacier was done at longitudinal profile. Ice and firn that cover the glacier are maximally 5 m thick (Figure 9). These two tends to melt during the ablation, being melted completely sometimes. The later led to direct heating and melting of the glacier's surface. Thin layer of rocky debris represents the boundary between the snow and firn with the glacier's ice. The surface of the glacier's ice is during the last two decades, at the end of ablation, more and longer exposed due to negative mass balance. As the consequence, the glacier's ice melts reducing its own thickness.

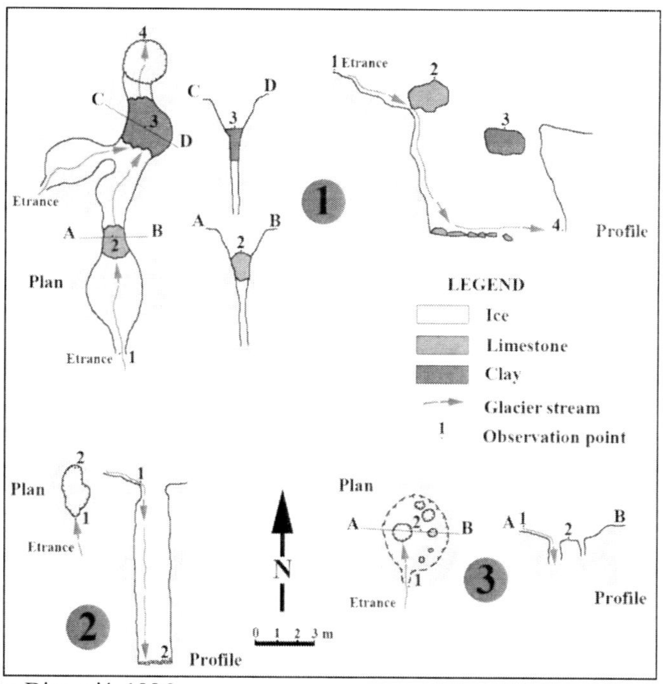

Modified by Djurović, 1996.

Figure 8. Glacial cave in the Debeli Namet glacier.

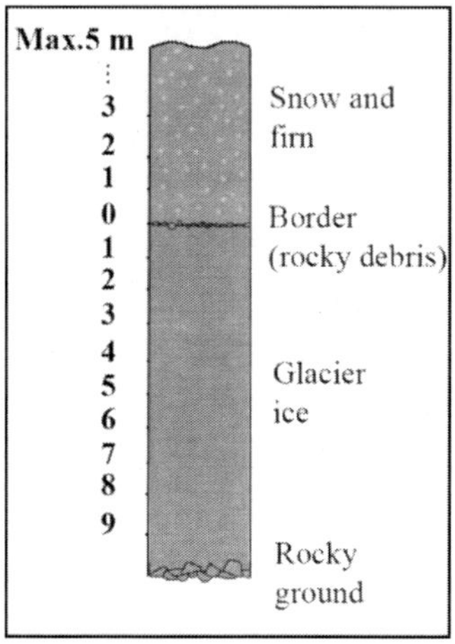

Figure 9. Profile across the Debeli Namet glacier.

During the recent researches (November 2015) was noted a gutter within the ice mass in the lowest part of the glacier tongue. It is approximately 50 m long and 1 m wide (Figure 7). The depth gradually raises leading to impassable cave near the end. Gutter owes its origin to a same cause as the caves themselves. Melt water flows over the glaciers surface and takes one direction. Having a higher temperature than the ice mass itself, it gradually melts the ice creating a gutter. This gutter enables the direct study of the structure of glacier tongue. On the sides of the gutter were noted sloping layers within the ice mass. Layers consist of very fine-grained material including smaller pieces of limestone debris, which formed through deposition on a surface of glaciers ice (Figure 7). The distance between two layers represents the annual growth of ice mass. Particularly interesting is the fact that dipping of layers does not coincide with the dip of glaciers mass, and is moreover quite opposite. Layers in gutter have a dip of 60°, while the glacier surface dips at some 20°. The most satisfactory reason for it is the change of inclination of bedrock. Downward the glacier starts to uplift and move towards terminal moraine, disturbing the dip of layers.

Glacier pedestals of various sizes are common at the glacier surface during the second half of the ablation period. Pedestals occur in medium part of the glacier, exactly at places where the largest amounts of debris accumulate. Such intensive accumulation is supported by numerous clefts and very severe movement of limestone debris at south sides of cirque.

One of the main problems is the origin and intensity of glacier feeding. Factors that led to glacier formation could be distinguished into two groups:

consistent and variable. The former are determined by orography: elevation, position in respect to relief forms and to direction of moisture currents movement, exposition, geological background, duration of expose to sun arrays, etc.

The most important variable factors are climate factors: regime, amount and kind of precipitation and the air temperature. The ratio between amount of snow that fall during the accumulation period and amount that will melt during the ablation period is the annual mass balance. It changes every year determining whether the glacier grows or recedes? Simply, the annual mass balance has the key role for changing in the glacier size.

The Debeli Namet glacier is located in the highest part of Pleistocene cirque Velika Kalica, where a recent secondary cirque has been formed. Recent cirque stretches north-east and is bordered from three sides by vertical, 400 m height cirque flanks. It is opened only to north-east (recent moraine ridge). Its bottom of 0.056 km^2 is also the maximal or theoretical aerial extent that the Debeli Namet glacier could reach without disruption of terminal moraine and its escape out of the cirque. This is the extent of the glacier at the beginning of the ablation period.

The drainage area for the Debeli Namet glacier is that of transporting snow to the cirque bottom, which takes part in the glacier feeding. It is about 0.2 km^2 and includes bottom and sides of the cirque. Annual amount of snow that directly precipitates at the bottom of cirque was calculated on account precipitation during months with negative means monthly air temperatures (November-April). In that period precipitates about 55% of estimated annual sum of 2600 m, what is 1430 mm. Sides of cirque, about 0.144 km^2 aerially, represent drainage area for avalanches. The amount of 3677 mm of snow brought by avalanches to the bottom was calculated multiplying spatial area of avalanches with annual snow amount and divided with the bottom extent. Adding this amount to those directly precipitated on the bottom gives the value of 5107 mm, i.e., the total amount of snow which feeds the glacier. At the beginning of the ablation period (neglecting amounts of wind-drifted or through evaporation lost snow) the average ice-firn mass is 286 000 000 m^3. It should be reduced nearly ten times at the end of ablation period.

The amount of wind-drifted snow (supply or loss) was not measured directly, though the precise estimation remained lacked. Geological background is supporting less intensive melting of the glacier. Melted water sink into limestone basement disabling the creation of pools and longer thermal influence to the ice mass.

In mid-nineties of the 20th century were performed exploration and sampling in order to determine the presence of heavy metals and radioactivity of ice from the Debeli Namet glacier (Veselinović et al., 1997). Studies were performed from the end of summer until the beginning of winter when is snow covering the thinnest. Borehole had 106 mm in diameter and was only 105 cm deep. At the end of the borehole only water and fine limestone substance occurred. This reflected that the borehole most probably reached the firn-glacier boundary as a melt-water percolates over it. It should be emphasized that drilling and sampling were performed in snow layer that accumulated until 1993.

Table 4. Concentration of Metals (μg/cm³) and Activity of Water and Ice (Bq/dm³) from a Borehole in the Debeli Namet Glacier (+ Determination hindered by other substances present)

Sample from depth cm	Pb	Cu	Zn	Na	K	Mg	Ca	Cd	K-40	Cs-134	Cs-137
0-10	2.3	5.1	+	+	+	+	+	<0.1	31.6	<0.524	<0.459
10-18	4.8	30.9	+	5210	660	710	37330	<0.1	14.1	<0.606	<0.525
18-26	1.2	22.9	+	2530	530	567	33693	<0.1	16.3	<0.493	<0.430
26-43	2.3	2.6	+	+	+	+	+	<0.1	7.4	<0.497	<0.466
43-55	9.6	8.2	+	+	+	+	+	<0.1	29.5	<0.824	<0.721
55	3.8	5.1	39.3	593	192	329	1965	<0.1	23.5	<0.994	<0.872
60	2.9	7.6	7.2	1037	208	307	27621	<0.1	30.9	<0.903	<0.819
70	10.5	30.0	6.0	1807	2160	724	14200	<0.1	25.1	<0.875	<0.804
Clean ice from 95	4.6	1.8	2.6	303	166	52	3050	<0.1	33.9	<0.441	<0.387
Unclean water from 100	1.86	1.7	+	+	+	+	+	<0.1	34.5	<0.451	<0.396
Water from borehole 5 min after driling from 105 cm	5.7	1.2	3.5	50	12	193	7520	<0.1	19.3	<0.271	<0.240
Water from borehole 15 min after driling from 105 cm	18.0	0.6	8.9	196	161	169	13350	<0.1	20.6	<0.260	<0.229
Water from borehole 30 min after driling from 105 cm	23.8	58.3	60.6	223	248	191	15520	<0.1	18.5	<0.278	<0.245
Clean, flowing water	10.6	0.9	12.4	148	90	915	8160	<0.1	13.4	<0.275	<0.244

Veselinović et al., 1997.

Table 5. Dynamic of the Debeli Namet glacier within the last 60 years

Year	Area km^2	Volume m^3	Source of data
August 1954	0.022	160 500	Djurović, 2012 – according to Belgrade Military Geographical Institute aerial photographs
August 1971	0.036	313 000	Djurović, 2012 – according to Belgrade Military Geographical Institute aerial photographs
August 1981	0.043	398 500	Djurović, 2012 – according to Belgrade Military Geographical Institute aerial photographs
August 7, 1993	0.017	113 000	Djurovic, 2012
August 5, 1995	0.031		Djurovic, unpublished
August 6, 1997	0.024	180 500	Djurovic, 2012
1998	0.012-0.014		Gachev, Stoyanov, 2012
2003	0.018	122 500	Hughes, 2008
2005	0.041	373 500	Hughes, 2008
October, 2006	0.050	489 000	Hughes, 2008
October 18, 2006	0.029	233 500	Djurovic, 2012
2007	0.037	325 000	Hughes, 2008
September 25, 2008	0.034	289 750	Djurovic, 2012
October 17, 2009	0.026	201 250	Djurovic, 2012
September 22, 2010	0.033	278 250	Djurovic, 2012
2011	0.027		Gachev, Stoyanov, Gikov, 2016
2012	0.020		Gachev, Stoyanov, Gikov, 2016
2013	0.030		Gachev, Stoyanov, Gikov, 2016
2014	0.024		Gachev, Stoyanov, Gikov, 2016
November, 2015	0.021		Djurovic, unpublished

Applying the method DPASV (A differential pulse anodic stripping voltammetry) was determined the presence of heavy metals Cu, Pb, Cd and Zn. Concentrations of K, Ca and Na was estimated using the flame emission spectrophotometry, while the concentration of Mg was recorded by atomic absorption spectrophotometry. Radioactivity was estimated by γ-ray spectrometry. Radioactive isotopes found in ice are assigned consequences from the explosion of "Lenin" Nuclear Power Plant at Chernobyl in 1986 (Table 4).

Introducing the dynamics of the Debeli Namet glacier is the extraordinary challenge and exploration problem. Implementation of different methods (remote sensing, direct field measurements, estimated values, etc.), along with different periods of observation (August, September, October and November) affect the accuracy of obtained results. The most precise results were obtained by direct measurements at the end of ablation period (October/November). Subjective factor should not be excluded as many authors took part in researches (Djurović, 2012; Hughes, 2007; 2008; Gachev et al., 2016). In spite a certain methodological discrepancy and subjective approaches the dynamic of the glacier within the last 60 years could be correctly accessed (Table 5).

The first data concerning the size of the glacier is based on the aerial photographs from 1954. The data for 1971 and 1981 was obtained by same methodology. The early first direct field observations of the glacier were conducted in 1993. Regular measurements by different authors have been applied since then, but institutional and systematic measurements still are lacking.

Although the data for each year within the last 61 years regarding the glacier size are missing, the average extent could be estimated. It is nearly 0.029 km^2 for the period from 1954 to 2015 (Figure 10). In 60% cases of the last ten years (since 2006) the glacier has been exceeding its average extent.

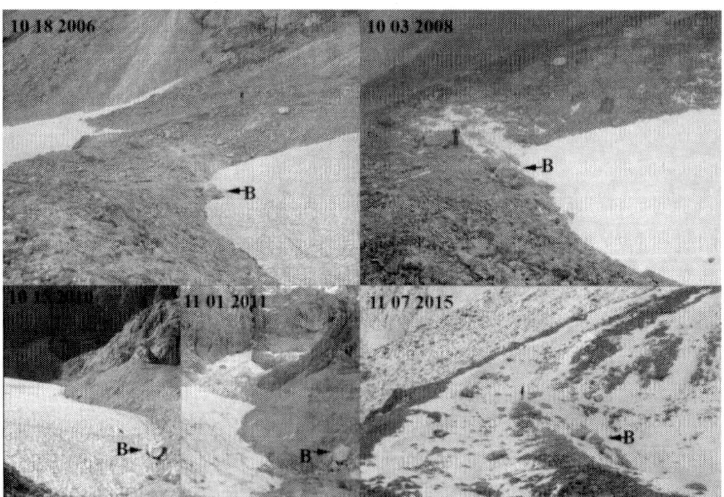

Figure 10. Different sizes of the Debeli Namet glacier. (Photographs by P. Djurović)

CONCLUSION

The Debeli Namet glacier is located on the south-easternmost part of Dinaric Mountain chain, on Durmitor Mountain in Montenegro. It occupies the highest elevation of the Pleistocene cirque Velika Kalica. The bottom of recent cirque being aerially 0.056 km^2 and with sides of 0.144 km^2 results in extent of 0.2 km^2. The glacier stretches from 2050 to 2180 m a.s.l. Its length is from 330 to 350 m, and average aerial extent 0.029 km^2. The average annual air temperature in the area where is the glacier is about 2.1°C, being during six months with negative men monthly temperature. Within the annual precipitation of 2600 mm the 1430 mm is a snow. The glacier feeds significantly beside directly from snowfalls (28%) by snow from avalanches (3677 mm), what makes 72%. These two gathered sources make 5107 mm. The biggest changes encounter the decrease in thickness, while the drop of length was less noticeable. The glacier had the largest aerial extent at the end of the 1970s and beginning of eighteens what coincides with the period of increased precipitation. The decreased precipitation rate and higher temperatures in certain years were compensated by broad drainage area of the glacier and resulted in insignificant omission of the glaciers size from its average.

REFERENCES

Coluccia, R. R., Guglielminb M. 2015. Precipitation–temperature changes and evolution of a small glacier in the southeastern European Alps during the last 90 years International. *Journal of Climatology* 35: 2783-2797.

Cvijić, J. 1889. Glacijalne i morfološke studije o planinama Bosne, Hercegovine i Crne Gore (Glacial and morphological studies on the mountains of Bosnia, Herzegovina and Montenegro), *Glas SKA* 57-21, Beograd.

Cvijić, J. 1903. Novi rezultati o glacijalnoj eposi Balkanskog poluostrva (The new results of the glacial epoch of the Balkan Peninsula), *Glas SKA LXV* 1/25, Beograd.

D'Alessandro, L., De Sisti, G., D'Orefice, M., Pecci, M., Ventura, R. 2003. Geomorphology of the Summit Area of the Gran Sasso d'Italia (Abruzzo Region, Italy). *Geografia Fisica e Dinamica Quaternaria* 26.

Djurović, P. Visokoplaninski kras Durmitora - geomorfološka studija (High mountainous karst Durmitor - geomorphological study). *Unpublished PhD thesis*, Geografski fakultet, Beograd, 1996.

Djurović, P. 1999. Natural and anthropogenic influences as threats to pits in the Debeli namet glacier Durmitor National park Montenegro. *Contribucion del estudio cientifico de las cavidades karsticas al conocimiento geologico Malaga* (Contribution of scientific study of the geological knowledge karst cavities Malaga), 579-585.

Djurović, P. 2002. New Approach in study of the extent of Pleistocene glaciation in the southwest part of Balkan peninsula. *International Scietific Conference in memory of prof.* Dimitar Jaranov Varna 2, Sofija.

Djurović, P. 2009. Reconstruction of the pleistocene glaciers of mount Durmitor in Montenegro. *Acta geographica Slovenica* vol, 49(2), 263-290.

Djurović, P. 2012. The Debeli Namet glacier from the second half of the 20th century to the present. *Acta geographica Slovenica* vol, 52(2), 277-294.

Djurović, P. Visokoplaninski kras Durmitora (High mountainous karst Durmitor). Univerzitet u Beogradu, Geografski fakultet, Beograd, Srb (University of Belgrade, Faculty of Geography, Belgrade, Serb), 2011.

Djurović, P. 2012. Comment on: "The glacial history of the Dinaric Alps, Montenegro" by P.D. Hughes, J.C. Woodward, P.C. van Calsteren, L.E. *Thomas Quaternary Science Reviews*, vol, 30, 3393-3412. *Quaternary Science Reviews* vol, 49, 23, 109-110.

D'Orefice, M., Pecci, M., Smiraglia, C., Ventura, R. 2000. Retreat of Mediterranean glaciers since the Little Ice Age: case study of Ghiacciaio del Calderone, Central Apennines, Italy. *Arctic, Antarctic and Alpine Research* 32, Colorado.

Dramis, F., Fazzini, M., Pecci, M., Smiraglia, C. 2002. The Geomorphological Effects of Global Warming in the Calderone Glacier Area (Central Apennine, Italy). In: *Extended Abstracts of the International Conference: Climate Changes, Active Tectonics and Related Geomorphic Effects in High Mountain Belts and Plateaux*. Addis Abeba.

Federici, P.R., Stefanini, M.C. 2001. Evidences and chronology of the Little Ice Age in the Argentera Massif Italian Maritime Alps. *Zeitschrift für Gletscherkunde und Glazialgeologie* (Journal of Glaciology and glacial geology) 37, Innsbruck.

Gachev, E. 2009. Indicators for modern and recent climate change in the highest mountain areas of Bulgaria. *Landform Analysis* 10, Poznań.

Gachev, E., Gikov, A., Zlatinova, C., Blagoev, B. 2009a. Present state of Bulgarian glacierets. *Landform Analysis* 11, Poznań.

Gachev E., Stoyanov K. 2012. Present day small perennial firn-ice patches in the mountais of the western Balkan peninsula Studia. *Geomorphologica Carpatho-Balcanica* vol, XLVI, 51-70.

Gachev E., Stoyanov K., Gikov, A. 2016. Small glaciers on the Balkan Peninsula: State and changes in the last several years. *Quaternary International*, Articles in Press.

Gabrovec, M., Hrvatin, M., Komac, B., Ortar, J., Pavšek, M., Topole, M., Triglav Čekada, M., Zorn, M. Triglavski ledenik (Triglav glacier), *Geografija Slovenije* (geography of Slovenia); Ljubljana, Sl, 2014, vol, 30.

Gams, I. 1994. Changes of the Triglav Glacier in the 1955-94 Period in the Light of Climatic Indicator. *Acta geographica Slovenica* vol, 34. Ljubljana.

Gómez, A., Palacios, D., Luengo, E., Tanarro, L.M., Schulte. L., Ramos, M. 2003. Talus instability in a recent deglaciation area and its relationship to buried ice and snow cover evolution (Picacho del Veleta, Sierra Nevada, Spain). *Geografiska Annaler* vol, 85A, Stockholm.

González-Trueba, J. 2004. Topoclimatical factors and very small glaciers in Atlantic mountain of SW Europe: the little ice age glacier advance in Picos de Europa (NW Spain). *Zeitschrift für Gletscherkunde und Glazialgeologie* (Journal of Glaciology and glacial geology) vol, 39, Innsbruck.

Grove, J. Little Ice Ages: Ancient and Modern. Routledge; London, UK, 2004.

Grünewald, K., Scheithauer, J., Gikov, A. 2008. Mikrolednik na planaina Pirin. *Problems of geography* 1-2: 159-174.

Grünewald, K., Scheithauer, J., 2008a. Bohrung in einen Mikrogletscher (Hole in a micro Glacier). *Zeitschrift für Gletscherkunde und Glazialgeologie* (Journal of Glaciology and glacial geology), 42(1): 3-18.

Grünewald, K., Scheithauer, J. 2010. Europe's southernmost glaciers: response and adaptation to climate change. *Journal of Glaciology* 56 (195): 129-142.

Hughes, P.D. 2007. Recent behaviour of the Debeli Namet glacier, Durmitor, Montenegro. *Earth Surface Processes and Landforms* 32: 1593-1602.

Hughes, P.D. 2008. Response of a Montenegro glacier to extreme summer heatwaves in 2003 and 2007. *Geografiska Annaler* vol, 90A (4), 259-267.

Hughes P.D., Woodward, J.C. Chapter 7: Glacial and Periglacial Environments. In: Woodward, J.C. (ed.) *The Physical Geography of the Mediterranean Basin.* Oxford University Press, Oxford, 2008.

Hughes, P.D., Woodward, J.C., Van Calsteren, P.C., Thomas, L. E. 2011. The glacial history of the Dinaric Alps, Montenegro. *Quaternary Science Reviews* vol, 30, 23-24, 3393-3412.

Jomelli V., Grancher D., Naveau P., Cooley D., Brunstein D. 2007. Assessment study of lichenometric methods for dating surfaces. *Geomorphology* vol, 86, 131-143. Doi: 10.1016/j.geomorph.2006.08.010.

Kern, Z., Surányi, G., Molnár, M., Nagy, B., Balogh, D. 2006. Investigation of natural perennialice deposits ice of Durmitor Mts, Montenegro. *Proceedings of the 2nd international Workshop on ice Caves*. Demänovská dolina.

Marović, M., Marković, M. 1972. Glacijalna morfologija šire oblasti Durmitora (Glacial morphology wider area of the Durmitor). *Geološki Anali Balkanskog poluostrva* (Geological Annals of the Balkan Peninsula) 37-2, Beograd.

Matthews, J.A. 2005. 'Little Ice Age' glacier variations in Jotunheim, southern Norway: a study in regionally controlled lichenometric dating of recessional moraines with implications for climate and lichen growth rates. *The Holocene* 15, 1-19.

Menković, M., Marković, M., Čupković, T., Pavlović, R., Trivic, B., Banjac, N. 2004. Glacial morphology of Serbian, with comments on the Pleistocene Glaciation of Monte Negro, Macedonia and Albania. *Quarternary Glaciations – Extent and Chronology*, Amsterdam.

Milivojević, M., Menković, Lj., Ćalić, J. 2008. Pleistocene glacial relief of the central part of Mt.Prokletije (Albanian Alps). *Quaternary International* 190, Amsterdam.

Milivojević, M. Glacijalni reljef na Volujku sa Biočem i Maglićem (Glacial relief on Volujak with Biocom and nebulae). *Geogrfaski institute "Jovan Cvijić" SANU* 68, Beograd, Srb, 2007.

Milojević, B. Ž. Visoke planine u našoj Kraljevini (High mountains in our Kingdom), Beograd, Srb, 1937.

Milojević, B.Ž. 1951. Durmitor – regionalno - geografska istraživanja (Durmitor - regional - geographical research). *Zbornik radova San* 9-2 (Proceedings San 9-2), Beograd.

Mirković, M. Geološki sastav i tektonski sklop Durmitora, Pivske planine i Volujka (Geologic structure and tectonic fabric of Durmitor, Piva Mountain and Volujka). *Posebno izdanje Geološkog glasnika Zavoda za geološka istraživanja SR Crne Gore* 5 (A special edition of the Geological messenger Institute of Geology SR Montenegro 5), Titograd, YUG 1983.

Nicod, J. 1968. Premières recherches de morphologie karstique dans le massif du Durmitor (First karst morphology of research in the Durmitor massif). *Meditérranée* vol 3, 187-216.

Pecci, M. 2006: Il Ghiacciaio del Calderone (Abruzzo settentrionale) (The Calderone Glacier (northern Abruzzo)). Testimoni Di Una Montagna Scomparsa (Witnesses Disappearance Of A Mountain). *Quaderno della Montagna* 8 (Notebook of Mountain 8).

Šifer, M. 1976. Nova dognanja na Triglavskem ledeniku in ledniku pod Skuto (Breakthroughs in the Triglav glacier and ledniku Skuta). *Acta geographica Slovenica* vol 15, Ljubljana.

Tintor, W. 1993. Die Kleingletscher der Julischen Alpen (The small glaciers of the Julian Alps). *Carinthia II* vol 183, 103, Klagenfurt.

Veselinović, D., Kovačević, D., Marković, D., Vukmirović, Z., Rajšić, S., Stanković, S., Stanković, A., Djordjević, D.1997. Sadržaj metala i radioaktivnost leda sa ostataka lednika Debeli namet na Durmitoru (The metal content and radioactivity ice with debris glacier Thick levy on Durmitor). *Zbornik radova: Rizik tehnoloških sistema i životna sredina* (Proceedings: The risk of technological systems and the environment), Niš, 145-148.

In: Glaciers
Editor: Nicholas Doyle
ISBN: 978-1-63484-941-8

Chapter 3

High Altitude Temperature Changes in the Tropical Andes Over the Last 15 000 Years Estimated from a Glaciological Model

***V. Jomelli*[1,*], *V. Favier*[2], *D. Brunstein*[2], *F. He*[3] *and Z. Liu*[4]**

[1]Université Paris 1 Pantheon-Sorbonne,
CNRS Laboratoire de Géographie Physique, Meudon, France
[2]Université Grenoble Alpes, LGGE CNRS, Grenoble, France,
[3]Center for Climatic Research, Nelson Institute for Environmental Studies, University of Wisconsin-Madison, Madison, Wisconsin, US
and College of Earth, Ocean, and Atmospheric Sciences, Oregon State University, Corvallis, OR, US
[4]Center for Climatic Research and Department of Atmospheric and Oceanic Sciences, University of Wisconsin-Madison, Madison, Wisconsin, US

Abstract

The warming trend during the deglaciation documented from polar ice core records was interrupted by cooler conditions during the Antarctic Cold Revearsal (ACR) in the Southern Hemisphere. Conversely

[*] Corresponding Author address: Email:vincent.jomelli@lgp.cnrs.fr.

temperature proxy records in the northern Hemisphere reveal two warm intervals during the ACR and the Holocene separated by the cold Younger Dryas (YD) event. The timing of tropical Andean glaciers fluctuations associated to these climate changes has been recently investigated. However little is known about regional paleoclimate conditions responsible for these glacier changes. In this, paper we estimate high altitude temperatures changes during the Late glacial and Holocene periods from a glaciological model. As a case study we focused on the Ritacuba Negro glacier located in the Colombian Cordillera de Cocuy. Based on the paleoglacial extents and within the range of uncertainties in precipitation changes, the modeling suggests that the ELA was at 4455 m als, 460 m below its current position about 14 kyr ago, which corresponds to a temperature 2.9° ± 0.8 °C lower than today. Subsequent glacier advances were reduced in size and became less important through time with a marked recession during the YD. At the end of the YD the ELA was at 4505 m asl, 50 m higher than during the ACR, due to a temperature increase of 0.3°C and at 4,635 m a.s.l. in the early Holocene corresponding to an additional warming of 0.8 °C, compared to the end of the YD. Our results suggest that the impact of future warming on glacier melting would exceed the retreat rate over the last 14,000 years. The Ritacuba Negro glacier should disappear by the end of the 21th century.

Keywords: tropical glacier, antarctic cold reversal, younger dryas, cosmogenic dating, positive degree day glaciological model, temperature reconstruction

INTRODUCTION

The climate during the deglaciation was marked by two abrupt and antiphased warm/cold climate periods. The warming trend in the Southern Hemisphere documented in polar ice core records was interrupted by cooler conditions during the Antarctic Cold Reversal (ACR ~14,500 to 12,800 years ago 12.8 kyr). Conversely temperature records from Greenland proxy records reveal two warm intervals during the ACR and the Holocene separated by the cold Younger Dryas (YD) event (~12,800 to 11,600 years ago). A recent study revealed that temperature decrease during the ACR was strongest in the South Atlantic and all regions south of 40°S (Pedro et al., 2015). At the same time, the terrestrial tropics and subtropics showed abrupt hydrologic variations that are significantly correlated with North Atlantic climate changes. Tropical glaciers are highly sensitive to changes in temperature and precipitation (Favier et al., 2004a,b; Rabatel et al., 2013), and accordingly, past fluctuations in glacier extent can be used to robustly reconstruct past changes in climate (Jomelli et al., 2009; Solomina et al., 2015). Temperature estimated from tropical glacier extent have been documented during the Holocene from different glaciological models and revealed significant changes. For instance in the northern tropical Andes of Venezuela, Polissar et al., (2006) estimated temperature declines of -3.2 ± 1.4°C and precipitation increases of 20% to retrieve the observed glacier changes during the last millennium. In the southern tropical Andes of Bolivia, Jomelli et al., (2011) calculated an annual mean temperature about -3.3 ± 0.8°C cooler 11,000 years ago and remained -2.1 ± 0.8°C cooler until the Little Ice Age (about 300 years ago). Associated to these temperature changes the authors estimated an equilibrium line altitude (ELA) depression of about 520 m during the early Holocene and about 300 m during the Little Ice Age (LIA) compared with today (Jomelli et al., 2011). In the present paper, we used the method proposed by Jomelli et al. (2011) to investigate the magnitude of atmospheric temperature and precipitation changes in the northern tropical Andes from the ACR-YD period which remains poorly constrained. To this end, our approach consisted in using a glaciological model forced by precipitation changes documented by a coupled global climate model to retrieve paleo glacier extents and ELA changes. These simulated paleo glacier fluctuations were constrained by an absolute moraine record.

Study Area

The study was conducted on the Ritacuba Negro glacier (2.8 km^2 in 2010; 5,271 m asl; 6°30′N, 72°20′W), located on the western slope of the Colombian Cordillera de Cocuy (Figure 1). Colombian glaciers represent about 4% of tropical glaciers surface (46,8 km^2 in 2010) and are currently retreating (Francou et al., 2013). They lost about 50% of their surface in the last 50 years. The Ritacuba Negro glacier is located in the ‘inner tropics’ (Kaser; 2001) with more or less continuous precipitation throughout the year. This region differs from the outer tropics where one wet season (when tropical conditions prevail from October to March) and one dry season (when subtropical conditions prevail from April to September) are observed. As illustrated on Figure 2, a very short drier season is observed from December to March while the maximum precipitation occurs from April to November. The wet season over the northern tropics is associated with the spatial shift of the Inter Tropical Convergence Zone (ITCZ) at its northern most position.

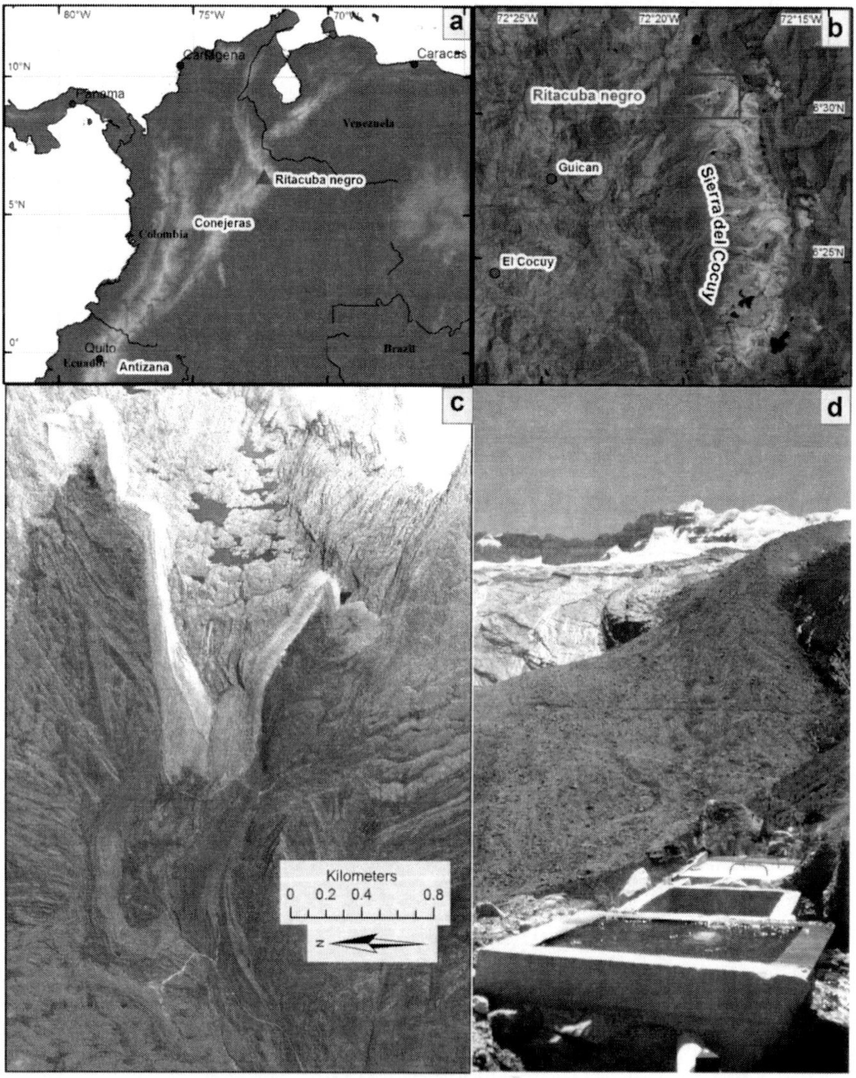

Figure 1. Location map of Ritacuba Negro glacier in the Sierra del Cocuy. a, Ritacuba negro and other glaciers analyzed in this study. b, Zoom on the location of Ritacuba Negro glacier. c, Map of the Ritacuba Negro glacier showing moraines considered in this study. d, a picture of Ritacuba Negro glacier showing the current front position and a lateral Holocene moraine.

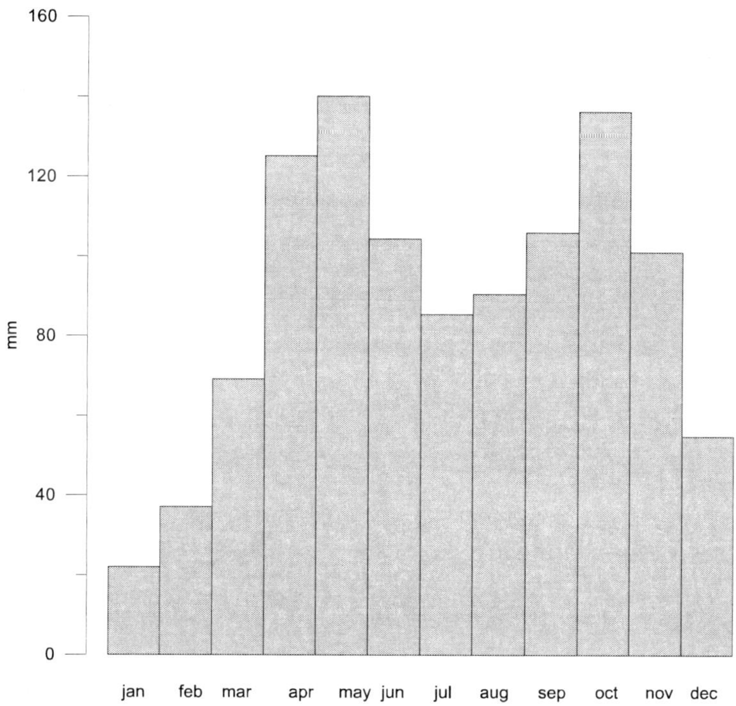

Figure 2. Precipitation in Ritacuba region (3716 m; 1975-2010).

METHODS

Glacier Chronology

In this study, we used three moraines to estimate paleo glacial extent of the Ritacuba Negro glacier. These moraines were selected from a pool of nine moraines previously dated by Jomelli et al., (2014). The first moraine selected in this study corresponds to the maximum Ritacuba Negro glacier extent in the last 14,500 years. The second moraine is located about 700 m upslope and corresponds to the YD period. The last moraine used in this study is located about 150 upslope and corresponds to the early Holocene. Each moraine ridge was mapped by means of a differential GPS survey in the field (accuracy <1m). These three moraines are composed of sandstone rock blocks. 15 large blocks (>60 cm high) were collected from near horizontal boulder top surfaces preferentially collected on stable parts of the moraine crest for cosmogenic dating preparation. Denudation was considered as absent for all of them. We used clinometer and compass to document the surrounding topographic shielding for all sample locations. All boulders were photographed and the ground-to-sample height of every boulder was measured. In situ-produced ^{10}Be was measured in the quartz mineral collected in the rock samples. Samples were crushed and sieved (250 μm – 1000 μm); magnetic components were eliminated by means of a Frantz magnetic separator. Quartz was extracted by dissolution of unwanted minerals in mixtures of HCl and H_2SiF_6. Solutions were renewed daily until they remained clear. Potential pollution by atmospheric ^{10}Be was removed by three successive leaching steps in HF, each step dissolving ~10% of the quartz. Before complete dissolution in HF (48%), a weighted amount (~100 mg) of a 3025 μg/g homemade ^{9}Be spike solution was added.

Successive separation by anion and cation exchange resins (DOWEX 1X8 then 50WX8) and precipitations were performed to isolate beryllium. The final hydroxide precipitates were dried and heated at 700°C to obtain BeO and finally mixed with niobium powder to prepare targets prior to their measurement by Accelerator Mass Spectrometer (AMS). All measurements were performed at the French AMS National Facility, ASTER, located at CEREGE in Aix-en-Provence. The data were calibrated directly against the National Institute of Standards and Technology (NIST) standard reference material 4325 by using an assigned value of $(2.79 \pm 0.03) \times 10^{-11}$ (Nishiizumi, et al., 2007). The chronology was established from the recent production rate of 3.9560.18 atoms g^{-1} yr^{-1} (Blard et al., 2013) with a time-dependent scaling and a specific Andes atmosphere model. Analytic uncertainties on the entire set of cosmic ray exposure ages averaged 6 ± 6%. The description of the samples and their age is given in Jomelli et al. (2014).

The Glaciological Model

To constrain potential combinations of precipitation and temperature changes that are consistent with the extents of the Ritacuba Negro glacier, we applied a model based on the well-established relationships between climate data and mass balance named the positive degree day (PDD) model (Braithwaite, 1985, 1995). The PDD model was run to estimate Equilibrium Line Altitude (ELA) and mass balance changes for the three distinct paleo-glacial extents documented in the field by moraine landscapes. It is based upon the notion that glacier melting occurs when air temperature 1–2 m above the glacier surface is above the melting point (0°C). The total melt over a period at some point is therefore proportional to the sum of positive temperatures at the same point, i.e., the positive degree day total. In the framework of long term climatic studies, it is possible to calculate monthly degree day totals from the monthly mean temperature, assuming that temperature is normally distributed within the month (Blard et al., 2007).

Melt is linked to positive degree-day total by the positive degree-day factors, with higher values for ice than for snow (Braithwaite, 1995). Therefore, there is greater melting if the surface material is ice compared with snow, under similar temperature conditions. The degree-day model is widely used for modelling the mass balance of mountain glaciers, following Braithwaite (1985). The degree-day model was successfully applied in the inner tropics to model current glacier melting (Maisincho, 2016; Maisincho et al., submitted). It can also be applied to reconstruct former glacier–climate conditions; (Blard et al., 2007; Jomelli et al., 2011). It provides a process-based approach utilizing detailed data on the relationship between glacier ablation and air temperature from modern glaciers.

Climate Model

The coupled atmosphere-ocean Global Circulation Model (AOGCM) employed in this study is the National Center for Atmospheric Research (NCAR) CCSM3 with a dynamic global vegetation module. CCSM3 is a global, coupled ocean–atmosphere–sea ice–land surface climate model without flux adjustment (Collins et al., 2006). All the simulations were performed in the version of T31_gx3v5 resolution (Yeager, et al., 2006). The atmospheric model is the Community Atmospheric Model 3 (CAM3) with 26 hybrid coordinate levels in the vertical and ~3.75-degree resolution in the horizontal. The land model uses the same resolution as the atmosphere, and each grid box includes a hierarchy of land units, soil columns, and plant types. Glaciers, lakes, wetlands and vegetated regions can be specified in the land units. The ocean model is the NCAR implementation of the Parallel Ocean Program (POP) in vertical *z*-coordinate with 25 levels. The longitudinal resolution is at nominal 3 degree and the latitudinal resolution is variable, with finer resolution near the equator (~0.9 degrees). The sea ice model is the NCAR Community Sea Ice Model (CSIM). CSIM is a dynamic-thermodynamic model that includes a subgrid-scale ice thickness distribution. The resolution of CSIM is identical to that of POP. The preindustrial control simulation reproduces the major features of global climate, notably in the deep ocean (Otto-Bliesner et al., 2006). The transient simulation was started at 22kyr from the end of an equilibrium run. The transient orbital forcing is from Berger (1978). The transient concentrations of the greenhouse gases (CO_2, CH_4 and N_2O) were adopted from Joss and Spahni (2008). The continental ice sheets were modified approximately once per thousand year according to the ICE-5G reconstruction (Peltier, 2004). The coastlines at LGM were also taken from the ICE-5G reconstruction and were changed twice at 12.9kyr and 6kyr. The meltwater flux was constructed roughly and follow the observations. More detailed discussions on the setup of this simulation can be found in Liu et al., (2009) and He (2011).

RESULTS

Moraine Chronology

The first moraine selected in this study corresponds to the maximum Ritacuba glacier extent in the late glacial-period. This outermost moraine of the Ritacuba sequence is located about 3.8 km from the current frontal position located at 4,660 m a.s.l. This moraine with frontal termination located at 3,975 m a.s.l, was dated to 13.9 ± 0.3 ^{10}Be kyr ago (n = 5) (Figure 3). The uncertainties associated with the ages account for analytical uncertainties (1 s.d.). Two samples dated to 15.2 ± 0.4 ^{10}Be kyr ago and 11.4 ± 0.9 ^{10}Be kyr ago, were considered as outliers on the basis of a χ^2 test reflecting isotope inheritance from previous exposure and post-depositional processes respectively. Including these two samples would have given a mean value of 14.1 ± 0.2^{10}Be kyr. The second moraine accumulation was formed at 11.8 ± 0.2 ^{10}Be kyr ago (n = 4) at the very end of the YD based on four internally consistent samples based on the χ^2 test. The last moraine used in this study was dated to 11.0 ± 0.4 ^{10}Be kyr ago (n = 4).

Glaciological Model Validation

Because Ritacuba glacier is located in the inner tropics (Kaser, 2001), the PDD model was first tested and calibrated on Antizana Glacier 15 (0°29′S,

78°8′W) (Favier et al., 2004) in Ecuador where 14 years surface mass balance data have been collected by GLACIOCLIM Observatory.

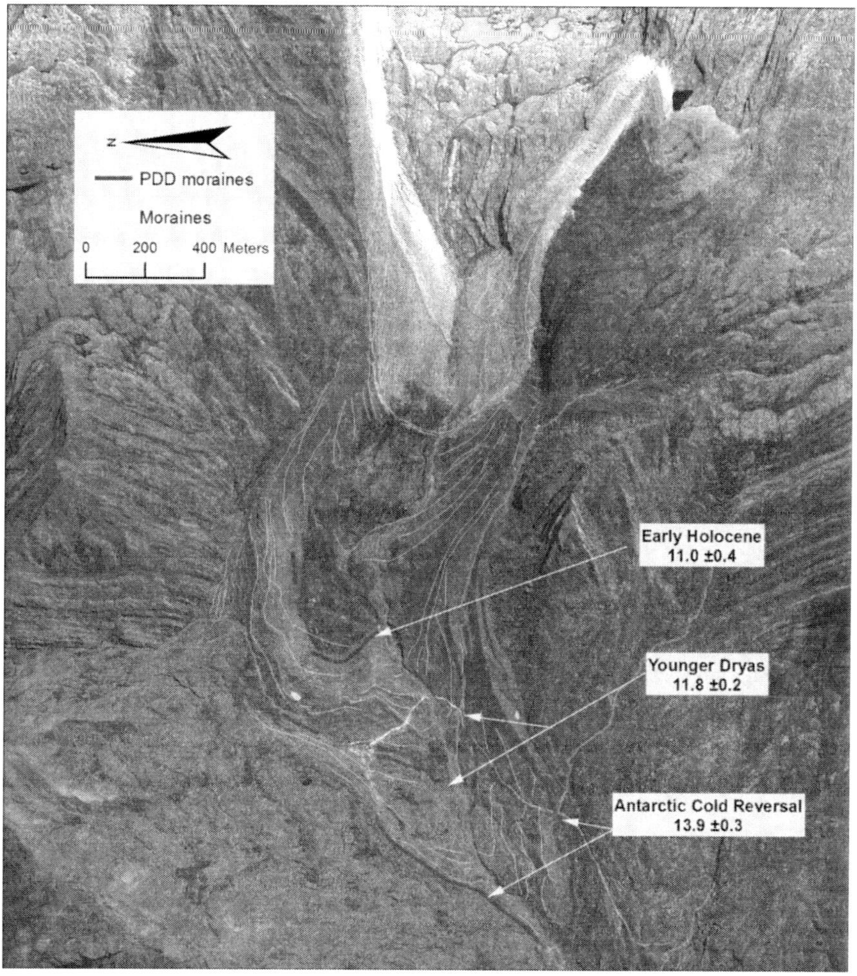

Figure 3. Map of the Ritacuba Negro glacier. Shown in red are dated moraines used in this study.

Antizana Glacier 15 has been extensively studied and is located under similar climatic settings, suggesting that glaciological processes are common to Ritacuba Negro glacier. Model calibration (Figure 4) gives different melt factors from those measured on Zongo Glacier, which is located in the outer tropics (Bolivia). This point is mainly linked to lower ablation and thermal seasonality and to a reduced ablation decrease with elevation in the inner tropics than in the outer tropics. The PDD model was next applied to Conejeras glacier (4° 49' N, 75° 22' W) (Figure 1) close to Ritacuba Negro glacier assuming melt factors obtained for Antizana Glacier 15. Calibration of PDD model on Conejeras glacier was based on mass balance measurements available between 2006 and 2011. The PDD model was applied to the present-day monthly temperature and precipitation measured at 3716 m asl in Ritacuba Valley leading to a perfect agreement with Conejeras glacier ablation measured from 4718 to 4889 m asl for 6 years between 2006 and 2011 (Figures 4). The PDD model was applied to Ritacuba Negro glacier to force the dynamics routine (cellular automata from Harper and Humphrey (2003)) yielding a glacier front at 4646 m asl, which is in perfect agreement with observation (4640 m asl), suggesting that the model is in good agreement with glaciological processes at Ritacuba Negro glacier.

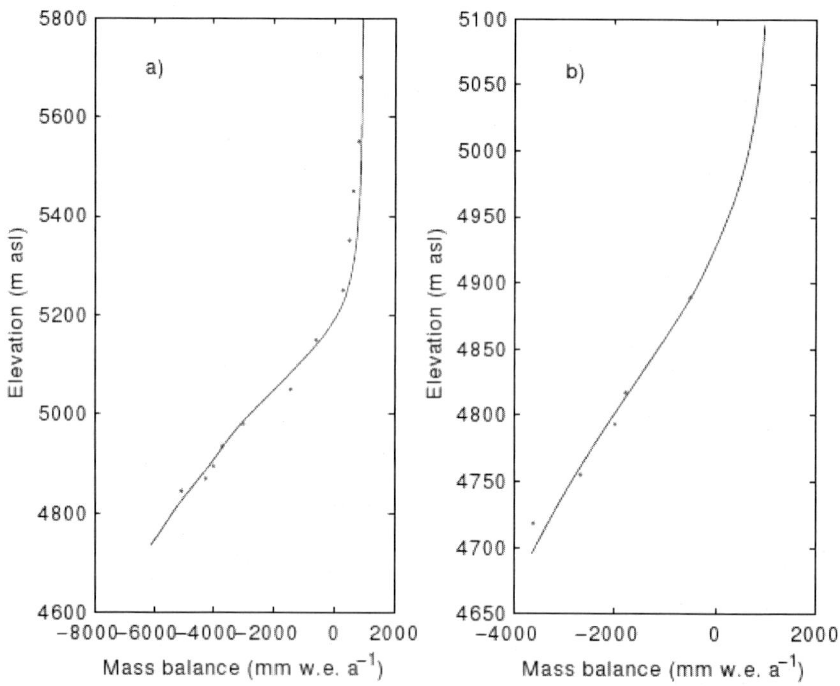

Figure 4. PDD Model results (blue line) versus mass balance measurements (red dots) a) for Antizana Glacier 15 (calibration step), b) and for Conejeras glacier (validation).

Estimation of Temperature Changes

To constrain potential combinations of precipitation and temperature changes that are consistent with the extents of the Ritacuba Negro glacier, we applied the PDD model to estimate Equilibrium Line Altitude (ELA) and mass balance changes for the three distinct paleo-glacial extents: (1) the ACR at 13.9 kyr (2) the very end of the YD at 11.8 kyr (3) and the early Holocene moraine at 11 kyr. Using the present-day climate data, the model yields an ELA at 4,915m a.s.l. (Figure 5) and a glacier front at 4,646 m a.s.l in good agreement with the currently observed front position. As there is a range of temperature/precipitation combinations with the PDD that can maintain one given glacier extent, independent estimates of precipitation were obtained from the transient CCSM3 model simulation of the last deglaciation (Liu et al., 2009; He 2011). Because glacier is controlled by regional climate conditions precipitation anomalies with respect to the ACR period were estimated in the Ritacuba region corresponding to 77-69°W, 2-10°N.

The CCSM3 simulation successfully represents the antiphased hemispheric temperature response to ocean circulation changes during the last deglaciation and, together with available global surface temperature proxy records, supports the conclusion that the superimposed global warming was driven by increasing carbon dioxide concentrations as inferred from polar ice cores (Liu et al., 2009; Shakun et al., 2012; He et al., 2013). Over the Ritacuba region, the CCSM3 simulation shows a 10%, 7% and 6% increase in annual precipitation as compared to modern conditions during the ACR, at the end of the YD and during the early Holocene respectively. During the main part of the YD however a 21% decrease in precipitation is simulated in agreement with proxy data (Haug et al., 2001). Based on the paleoglacial extents and within the range of uncertainties in precipitation changes, the PDD modeling suggests that the ELA was at 4,455m a.s.l at 13.9 kyr, i.e., 460 m below its current position, which corresponds to a temperature 2.9° ± 0.8°C lower than today (Figure 5). At the end of the YD the ELA was 50 m higher than during the ACR, due to a temperature increase of 0.3°C (i.e., 2.6° ± 0.8°C lower than today) and at 4,635 m a.s.l. in the early Holocene corresponding to an additional warming of 0.8°C, compared to the end of the YD (i.e., 1.8°C° ± 0.8°C cooler than today). Such pattern is qualitatively consistent with temperature estimated by the model. A -2.5°C lower than today is estimated at 13.9 kyr but the model shows much warmer temperatures at the YD and early Holocene periods with values similar as today.

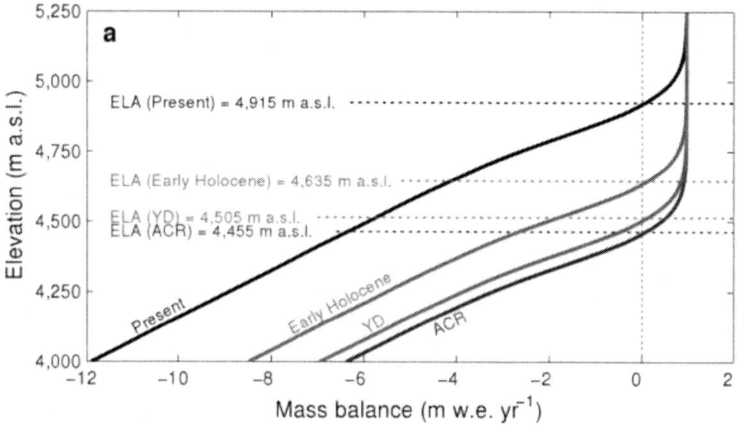

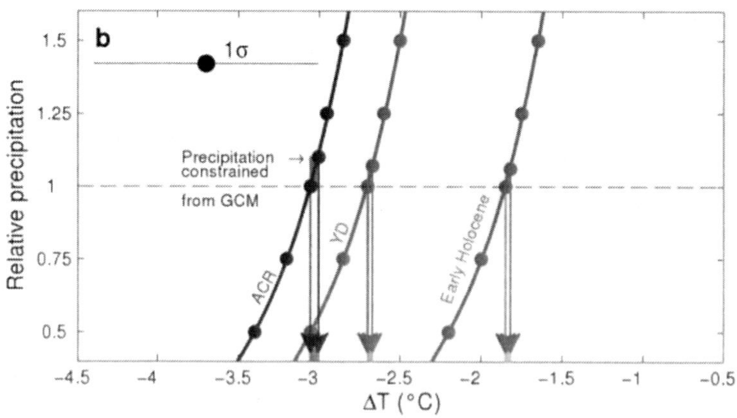

Figure 5. Palaeoclimatic reconstructions of the Ritacuba negro glacier. a, Plot of mass balance against elevation for the present day (black curve, Present), 13.1 kyr ago (blue, ACR), 11.6 kyr ago (red, YD) and 10.6 kyr ago (green, Early Holocene). Dashed lines are altitudes of the ELA from 13.1 kyr ago to present. The present-day mass balance curve at the Ritacuba negro glacier is characterized by an ELA at 4,915 m a.s.l. b, Palaeoclimatic conditions required to reproduce Ritacuba glacier extent during the past 13.6 kyr. Changes in temperature (ΔT) are plotted against relative changes in precipitation. The 1σ error bar shows the temperature uncertainty arising from the model assumptions. Sensitivity tests indicate that the temperature uncertainty is ± 0.7°C. Relative precipitation is normalized to present day value. Vertical lines with coloured arrows indicate the range of possible temperature depression for 13.1 kyr ago (blue), 11.6 kyr ago (red), 10.6 kyr ago (green) under a scenario of precipitation increase as suggested by our GCM output, and a scenario of constant precipitation regarding to current observed situation.

DISCUSSION AND CONCLUSION

Our paleoclimate record from Ritacuba Negro glacier helped us to understand the response of the tropical glaciers and regional oceanic and atmospheric circulation systems to natural climate variability in order to place recent and future changes in a longer-term perspective. Several studies have shown a significant warming trend in the tropical Andes over the last decades. For instance Vuille et al. (2008a) showed that near-surface air temperature increased significantly (by 0.10°C per decade) in the last 70 yr, which represents an overall temperature increase of 0.68°C since 1939. Future projections estimate a 4–5°C warming in the high mountains of the tropical Andes by 2100 using a regional climate model (RCM) based on two different A2 and B2 emission scenarios (Bradley et al., 2006; Urrutias and Vuille, 2009). This warming would be enhanced at higher elevations than 4000 ma.s.l and further amplified in the middle and upper troposphere. Under this warming condition glaciers are expected to present higher ablation in the future.

However, changes in precipitation may also occur and impact glacier surfaces. Several precipitation changes have been documented in the past. Precipitation has slightly increased in the second half of the 20th century in the inner tropics and decreased in the outer tropics (Vuille et al., 2008a), but on a longer time scale changes in precipitation are not clear in the Andes, due to the lack of long-term continuous records. Projected changes in precipitation are spatially much less coherent than temperature, and models exhibit both regions with increasing or with decreasing precipitation across the Andes. In the Ritacuba Negro region, changes in the seasonality are expected, with higher precipitation during the core of the wet season but lower during the dry period. This would effectively enhance the seasonal hydrological cycle in the Ritacuba region. Our glacier model suggests that the impact of future warming will exceed the one from increasing precipitation during the wet season. Consequently one can expect that glacier retreat at the end of the 21th century will exceed the retreat documented in the past 14,000 years. The front of Ritacuba glacier retreated to 3.8 km over the last 14,000 years but the glacier will very likely disappear at the end of the 21th century. Many other small, low-lying glaciers are already completely out of equilibrium with current climate and will disappear within a few decades. Assuming that longer and stronger dry periods are expected in the future, this retreat will have important impact on high elevation water resources and on periods with low discharge.

ACKNOWLEDGMENTS

Financial support was provided by the French ANR El Paso programme no. 10-blan-68-01 and LMI GREAT ICE. D. Bourles, G. Aumaître and K. Keddadouche are thanked for their valuable assistance during the measurements performed at the ASTER AMS national facility (CEREGE, Aix en Provence) which is supported by the INSU/CNRS, the ANR through the "Projets thématiques d'excellence" program for the "Equipements d'excellence" ASTER-CEREGE action, IRD and CEA. We thank the staff at SARM-CRPG for the compositional analysis. F.H. was supported by the US NSF (AGS-1203430) and by the NOAA Climate and Global Change Postdoctoral Fellowship program, administered by the University Corporation for Atmospheric Research. This research used resources of the Oak Ridge Leadership Computing Facility at the Oak Ridge National Laboratory, which is supported by the Office of Science of the U.S. Department of Energy under Contract No. DE-AC05-00OR22725

REFERENCES

Berger, A.L. (1978). Long-Term Variations of Daily Insolation and Quaternary Climatic Changes. *Journal of the Atmospheric Sciences* 35, 2362-2367.

Blard, P.H., Lavé, J., Pick, R., Wagnon, P., Bourles, D. (2007). Persistence of full glacial conditions in the central Pacific until 15.000 years ago. *Nature* 449, 591-594.

Blard, P. H., Braucher, R., Lave, J. Bourles, D. (2013).Cosmogenic 10Be production rate calibrated against 3He in the high tropical Andes (3800–4900 m, 20–22°S). *Earth Planet. Sci. Lett.* 382, 140–149.

Bradley, R. S., Vuille, M., Diaz, H., Vergara, W. (2006). Threats to water supplies in the tropical Andes. *Science* 312, 1755 – 1756, doi:10.1126/science.1128087

Braithwaite, R.J. (1995). Positive degree day factors for ablation on the Greenland ice sheet studied by energy balance modelling. *Journal of Glaciology* 41, 153-160.

Braithwaite, R.J. (1985). Calculation of degree-days for glacier-climate research. *Zeitschrift für Gletscherkunde and Glazialgeologie,* 20, 1-8.

Collins, W., Bitz, C.M., Blackmon, M.L., Bonan, G.B., Bretherton, C.S., Carton, J.A., Chang, P., Doney, S.C., Hack, J.J., Henderson, T.B., Kiehl, J.T., Large, W.G., McKenna, D.S., Santer, B.D., Smith, R.D. (2006).The Community Climate System Model Version 3 (CCSM3). *J. Climate* 19, 2122-2143.

Favier, V., Wagnon, P., Chazarin, J.-P. Maisincho, L., Coudrain, A. (2004a). One-year measurements of surface heat budget on the ablation zone of Antizana Glacier 15, Ecuadorian Andes. *J. Geophys. Res.* 109, D18105, doi:10.1029/2003JD004359.

Favier, V., Wagnon, P., Ribstein, P. (2004b). Glaciers of the outer and inner tropics: a different behavior but a common response to climatic forcing. *Geophys. Res. Lett.* 31, doi:10.1029/2004GL020654.

Francou, B., Rabatel, A., Soruco, A., Sicart, J.E., Silvestre, E., Ginot, P., Cáceres, B., Condom, T., Villacís, M., Ceballos, J.L., Lehmann, B., Anthelme, F., Dangles, O., Gomez, J., Favier, V., Maisincho, L., Jomelli, V., Vuille, M., Wagnon, P., Lejeune, Y., Ramallo, C., Mendoza, J. (2013). Glaciares de los Andes tropicales, victimas del cambio climatico. CAN – PRAA – IRD. National Lib Lima. 96pp. ISBN 978-9972-787-26-7.

Harper, J. T. and Humphrey, N. F. (2003). High altitude Himalayan climate inferred from glacial ice flux. *Geophys. Res. Lett.* 30, 1764–1767.

Haug, G. H., Hughen, K. A., Sigman, D. M., Peterson, L. C., Röhl, U. (2001). Southward Migration of the Intertropical Convergence Zone through the Holocene. *Science* 293, 1304-1308.

He, F. Simulating transient climate evolution of the last deglaciation with CCSM3. (2011). Ph.D thesis, Dept. Atmos. and Oceanic Sci., Univ. Wisconsin-Madison. 161 pp.

He, F., Shakun, J.D., Clark, P.U., Carlson, A.E., Liu, Z., Otto-Bliesner, B.L., and Kutzbach, J.E. (2013). Northern Hemisphere forcing of Southern Hemisphere climate during the last deglaciation. *Nature* 494, 81-85.

Jomelli, V., Favier, V., Vuille, M., Braucher, R., Blard, P-H., Khodri, M., Colose, C., Brunstein, D., Bourlès, D., Leanni, L., Rinterknecht, V., Grancher, D., Francou, B., He, F., Ceballos, J.L., Francesca, H., Liu, Z., Otto-Bliesner, B. A major advance of tropical Andean glaciers during the Antarctic Cold Reaversal (2014). *Nature* 513,224–228 doi:10.1038/nature13546.

Jomelli, V., Khodri, M., Favier, V., Brunstein, D., Ledru, M.P., Wagnon, P., Blard, P.H., Sicart, J.E., Braucher, R., Grancher, D., Bourlès, D., Braconnot, P., Vuille, M. (2011). Irregular tropical glacial retreat during the Holocene driven by progressive warming. *Nature* 474, 196-199.

Jomelli, V., Favier, V., Rabatel, A., Hoffmann, G., Brunstein, D., Francou, B. (2009). Fluctuations of Andean tropical glaciers since the last millennium and palaeoclimatic implication: a review. *Palaleogeography, Palaleo climatology Palaleoecology* 281, 269-282.

Joos, F., Spahni, R. (2008). Rates of change in natural and anthropogenic radiative forcing over the past 20,000 years. *Proc Nat Aca Sci.* 105, 1425-1430, doi:10.1073/pnas.0707386105.

Kaser, G. (2001). Glacier–climate interactions in low latitudes. *J. Glaciol.* 47, 195- 204.

Liu, Z., Otto-Bliesner, B.L., He, F., Brady, E.C., Tomas, R., Clark, P.U. Carlson, A.E., Lynch-Stieglitz, J., Curry, W., Brook, E., Erickson, D., Jacob, R., Kutzbach, J., Cheng, J. (2009). Transient simulation of deglacial Climate Evolution with a new mechanism for Bolling-Allerod warming. *Science* 325, 310-314.

Maisincho, L., Favier, V., Wagnon, P., Francou, B., Villasis, M., Basantes, R., Mourre, L., Jomelli, V., Rabatel, A., Caceares, B. (submitted). Mass balance modeling in the inner tropics using a positive degree-day approach.

Maisincho, L. (2016). Analyse de la fonte glaciaire et nivale dans les Andes tropicales à partir d'un bilan d'énergie: Glacier de l'Antisana, Equateur (0°28'S), PhD thesis, Université Grenoble Alpes, 197 pp.

Nishiizumi, K., Imamura, M., Caffee, M., Southon, J., Finkel, R., McAnich. J. (2007). Absolute calibration of ^{10}Be AMS standards. *Nuclear Instruments and Methods in Physics Research Section B: Beam Interactions with Materials and Atoms* 258, 403-413.

Otto-Bliesner B. L. Brady, E.C., Clauzet, C., Tomas, R., Levis, S., Kothavala, Z. (2006). Last Glacial Maximum and Holocene Climate in CCSM3. *J. Clim.* 19, 2526-2544.

Pedro, J.B., Bostock, H.C., Bitz, C.M., He, F., Vandergoes, M.J., Steig, E.J. Chase, B.M., Krause, C.E., Rasmussen, S.O., Markle B.R., Cortese, G. (2015). The spatial extent and dynamics of the Antarctic Cold Reversal. *Nature Geoscience* | DOI: 10.1038/NGEO2580.

Peltier, W. R. (2004). Global Glacial Isostasy and the Surface of the Ice-Age. *Annu. Rev. Earth Planet. Sci.* 32, 111-149.

Polissar, P.J. Abbott, M.B., Wolfe, A.P., Bezada, M., Rull, V., Bradley, R.S., 2006. Solar modulation of Little Ice Age climate in the tropical Andes. *PNAS* 103, 8937–8942.

Rabatel, A., Francou, B., Soruco, A., Jesus Gomez, J., Cáceres, B., Ceballos, J.L., Basantes, R., Vuille, M., Sicart, J.E., Huggel, C., Scheel, M.,

Lejeune, Y., Arnaud, Y., Collet, M., Condom, T., Consoli, G., Favier, V., Jomelli, V., Remigio Galarraga, R., Ginot, P., Maisincho, L., Mendoza, J., Ménégoz, M., Ramirez, E., Ribstein, P., Suarez, W., Villacis, W., Wagnon, P. (2013). Current state of glaciers in the tropical Andes. A multi secular perspective on glacier evolution and climate change. *The Cryosphere* 7, 81-102.

Shakun, J.D., Clark, P.U., He, F., Marcott, S.A., Mix, A.C., Liu, Z., Otto-Bliesner, B.,Schmittner A., Bard, E. (2012).Global warming preceded by increasing carbon dioxide concentrations during the last deglaciation. *Nature* 484, 49-53.

Solomina, O., Bradley, R.S., Hodgson, D., Ivy-Ochs, S., Jomelli, V., Mackintosh, A., Nesje, A., Owen, L.A., Wanner, H., Wiles, G., Young N.E. (2015). Holocene glacier fluctuations. *QSR*, 111, 9-34.

Urrutia, R., Vuille, M. (2009). Climate change projections for the tropical Andes using a regional climate model: temperature and precipitation simulations for the end of the 21st century. *J. Geophys. Res*. 114, D02108, doi:10.1029/2008JD011021.

Vuille, M., Francou, B., Wagnon, P., Juen, I., Kaser, G., Mark, B., Bradley, R. (2008). Climate change and tropical Andean glaciers: past, present and future. *Earth Sci. Rev*. 89, 79–96.

Yeager, S.G., Shields, C.A., Large, W., Hack, J. (2006).The Low-Resolution CCSM3. *J. Climate* 19, 2545-2566.

In: Glaciers
Editor: Nicholas Doyle

ISBN: 978-1-63484-941-8

Chapter 4

THE EFFECT OF CLIMATE CHANGE ON MICROBIAL COMMUNITIES FROM GLACIERS

Eva García-López, Paula Alcázar, Marina Postigo and Cristina Cid*

Microbial Evolution Laboratory, Center for Astrobiology (CSIC-INTA), Torrejón de Ardoz, Spain

ABSTRACT

Glaciers have been successfully colonized by numerous microorganisms; including representatives of bacteria, archaea, unicellular eukaryotes and viruses. In recent decades, the most pronounced effects of climate change have been observed in glaciers. The microorganisms in these regions are specially adapted to extreme environments, in which they live. They can be highly vulnerable to changes in climate. Ecosystems considered extreme often combine different environmental factors that are thought to be incompatible with most forms of life. They are highly selective and confined habitats, yet favorable for adaptation and speciation. These regions are not only rich in biodiversity, but also in new mechanisms of adaptation to the

---

* Corresponding author: Dr. Cristina Cid, Microbial Evolution Laboratory, Center for Astrobiology (CSIC-INTA), Ctra. Ajalvir, km 4, 28850 Torrejón de Ardoz, Madrid, Spain. E-mail: cidsc@inta.es, fax: +34915201074.

environment, since the selection acts with a special intensity. These features make them unique and interesting for the study and protection of the biological heritage. They are also extremely fragile areas against climatic or environmental changes.

The vast majority (90-99%) of these organisms cannot be recovered as axenic cultures under standard laboratory conditions, and therefore are not likely to be studied by classical methods. However, thanks to DNA sequencing methods, a lot of information about their biodiversity has been acquired. This chapter summarizes the current knowledge of the biodiversity and ecology of microbial communities from glaciers, including our latest results on this topic.

Abbreviations and Terminology

OTUs	operational taxonomic units
16S rRNA	a large polynucleotide (~1500 bases) that functions as a part of the small subunit of the ribosome of prokaryotes (Bacteria and Archaea) and from whose sequence evolutionary relationships can be obtained
18S rRNA	eukaryotic counterpart of 16S rRNA
HSPs	Heat Shock Proteins

Keywords: extremophile, psychrophile, diversity, ecology, glacier, Svalbard, Antarctica

1. Introduction

At the present time, glaciers, which include mountain glaciers, ice caps and ice sheets, cover about 10% of the Earth's land surface (Hambrey and Alean, 2004). These glaciers have been successfully colonized by numerous microorganisms; including representatives of bacteria, archaea, unicellular eukaryotes and viruses. The discovery of cold-tolerant microorganisms in permanently frozen environments has broadened the known range of environmental conditions which support microbial life.

In recent decades, the most pronounced effects of climate change have been observed in glaciers. The microorganisms in these regions are specially adapted to extreme environments, in which they live. They can be highly vulnerable to changes in climate. Glacial ecosystems are considered extreme, and often combine different environmental factors that are thought to be incompatible with most forms of life; these mainly consist of cold temperatures, desiccation, high irradiation including ultraviolet, low nutrients and photoreactive chemistry. Glaciers are highly selective and confined habitats, which make them favorable environment for adaptation and speciation. These regions are not only rich in biodiversity, but also in new mechanisms of adaptation to the environment; since selection acts with a special intensity. These features make them unique and interesting for the study and protection of the biological heritage. In addition, these microorganisms are of great interest in biotechnology (Garcia-Descalzo et al., 2013).

2. Microorganisms in Glaciers

Glaciers are inhabited by microorganisms belonging to all domains of life, generally named psychrophiles and psychrotolerant or psychrotroph. Since 1902, when Schmidt-Nielsen first described a bacterium able to grow and survive at 0°C (Schmidt-Nielsen, 1902; Russell, 1992), the scientific community has for years kept an open debate (especially in the 60s) on the more or less proper use of the term psychrophile, which experienced periods of confusion with other nearby terms such as psychrotolerant or psychrotroph. In order to clarify the situation and to eliminate future confusion Richard Y. Morita published a review in 1975 (Morita, 1975) proposing the definition of psychrophile as an organism with an optimum growth temperature around 15°C or less, a maximum temperature of growth around 20°C and a minimum growth temperature at 0°C or below. In this line, the psychrotolerant or psychrotrophic microorganisms are defined as those that maintain the ability to grow at 0°C but especially close to this point (3°C-5°C) with an optimal growth temperature above 15°C, usually around 20°C, and the upper limit above 20°C; in some cases up to 40°C (Barros et al., 1978; Russell, 1990). Although the distinction between psychrophiles and psychrotolerant was more clearly set from a paper published by Russel in 1993 based on biochemical differences (Russell, 1993).

Psychrophiles are typically found in permanent cold environments (Russell, 1992), where they are often the dominant type (Wynn-Williams, 1990); and mainly in aquatic environments where the temperature is more stable (Russel, 1990). They endure narrower temperature ranges than psychrotrophs. Psychrotrophs are characteristic of cold habitats in which the temperature fluctuates both throughout the day and throughout the year, being therefore very adaptable and versatile (Russell et al., 1998).

However, other authors such as D'Amico (D'Amico et al., 2006) encompassed in the term psychrophile all those organisms with the ability to grow well at temperatures close to the freezing point of water. At temperatures near - 20ºC, bacteria in sea ice and permafrost have been found. They maintain metabolic activity due to the existence of a small amount of unfrozen water inside the permafrost, or ice with a high salt concentration (D'Amico et al., 2006).

The ability of glacier microorganisms to survive at low temperatures implies that they have overcome key barriers inherent to permanently cold environments (Alcazar et al., 2009; Anesio et al., 2012). They use different strategies to cope with low temperatures (De Maayer et al., 2014; Maccario et al., 2015): increased membrane fluidity, freeze protection, synthesis of cold-adapted enzymes, changes in protein-protein interactions (Garcia-Descalzo et al., 2011), maintenance of catalysis at cold temperatures, and synthesis of pigments (Martin-Cerezo et al., 2015).

Glacial microorganisms maintain their metabolisms at temperatures below 0ºC, not only by photoautotrophy, where sunlight penetrates the ice but also by chemoautothrophy in dark regions of glacial ice.

Frozen water contained in glacial ice is not available for cellular processes, but most icy habitats contain some liquid water, at least temporally. It has been shown (Price, 2000; Mader et al., 2006) that the structure and chemical composition of these microenvironments are in the range typically used in laboratory media, and can allow bacterial survival. Another microbial habitat, which has been widely studied in Arctic, Antarctic and Alpine glaciers, is constituted by the cryoconite holes (Cameron et al., 2011; Edwards et al., 2011). They are melted little holes on the ice surface filled with water and particles of debris that absorb more solar radiation than the surrounding ice. Two main differences have been described between these two glacier habitats: glacier ice and cryoconite holes (Miteva, 2008). Firstly, microbial abundances and diversity are higher in cryoconite holes than in glacier ice. Secondly, photosynthetic bacteria act as an important source of organic carbon needed for the rest of the microbial community in cryoconite holes.

Glacial habitats contain small number of cells compared to other sites (Boetius et al., 2015, 681). Comparisons of geographically distinct glaciers worldwide have shown a great variation in microbial biomass and community structure (Miteva et al., 2004; Zhang et al., 2009; Garcia-Descalzo et al., 2013).

Regarding Prokaryotes, there have been a number of studies on the microbiological composition of such psychrophilic populations in recent years. Viable bacterial communities have been observed beneath glaciers, both in the northern (Skidmore et al., 2000) and southern hemispheres (Foght et al., 2004). Cyanobacteria are the dominant phototrophs across the glacial surfaces. They have frequently been reported from geographically different glaciers in Arctic (Stibal et al., 2006) and Antarctic (Taton et al., 2003) areas. Other bacterial classes such as Proteobacteria and Bacteroidetes are also very frequent in glaciers (Xiang et al., 2009).

Concerning microeukaryotes, several studies have analyzed the diversity of small eukaryotes, sampled in different cold ecosystems and studied by gene cloning and sequencing of rRNA genes. They have shown high phylogenetic diversity (Lefranc et al., 2005). These studies identified a wide variety of lineages and retrieved sequences not clearly assigned to any known organisms. It should be emphasized that most of these studies on small eukaryote diversity were conducted in marine and lake ecosystems (Medinger et al., 2010; Bachy et al., 2011). Thus, little is known about the diversity of eukaryotic microorganisms from glaciers. De García et al. (2007) described the occurrence of viable yeasts in the meltwaters running off glaciers, in the northwest Patagonia. Additionally, Butinar et al. (2007) isolated culturable yeasts from basal ice layers of high arctic glaciers in the Svalbard Islands. Likewise, Margesin et al. (2007) and Turchetti et al. (2008) described the existence of yeasts in some Alpine glacier habitats (e.g., cryoconites, ice cores, sediments and meltwaters); and Branda et al. (2010) in Calderone Glacier, Apennines. Other groups of protists such as Metazoa, Rhizaria, Viridiplantae and Stramenopiles were identified by Garcia-Descalzo et al., (2013), in the Pyrenean glaciers.

3. The Effect of Climate Change on Glaciers

In recent years, most climatologists agree that human activities are changing the Earth's climate (Immerzeel et al., 2010). This is partly due to the increasing levels of greenhouse gases, particularly carbon dioxide, from the burning of fossil fuels. Other atmospheric gas with increased levels, methane, is attributable to the melting of permafrost and the increasing number of domestic animals.

According to the Intergovernmental Planet on Climate Change reports (IPCC, 2014), if additional efforts are not made to reduce emissions of greenhouse gases other than the currently being made, emissions are expected to persist, driven by the growth of the world's population and economic activities. Increases in global mean surface temperature in 2100 will range from 3.7°C to 4.8°C above the 1850-1900 average temperature.

In the period between 1992 and 2011, the ice sheets of Greenland and Antarctica have lost mass, and that loss has occurred at a faster race between 2002 and 2011. Glaciers have continued declining in most of the world. Projections suggest that the overall volume of glaciers, excluding the periphery of Antarctica (and excluding the ice sheets of Greenland and Antarctica) decrease between 15% and 55% in the best scenario, and between 35% and 85%, in the worst and more serious case (IPCC, 2014).

Signs of global warming in the form of glacier recession are present in most parts of the world (Figure 1). A possible consequence of permafrost thawing, based upon predicted global warming scenarios, is that there may be an increase in microbial activity, and an increase on the active layer thickness (Gilichinsky et al., 2006). Thus, the current atmospheric warming will have severe consequences for the structure and functioning of glacier ecosystems, with changes that, in turn, may feedback on the global-scale composition of the atmosphere (Michelsen et al., 2012; Wagner et al., 2012).

4. The Effect of Climate Change on the Microbial Communities of Glaciers

It is well known that worldwide glaciers are strongly retreating due to the ongoing climate change. Glacial ice is a unique matrix capable of protecting microorganisms for long-term preservation. The organisms entrapped in glacial ice can provide information on evolutionary processes and ancient biodiversity (Ma et al., 2000). Cold temperatures are a driving force in the diversification and evolution of microorganisms. Glacial ice cores provide unequalled records of climatic evolution on a timescale of millions of years. Furthermore, the effect of global warming on microorganisms from glaciers is relevant since they constitute the base of the food webs and have been reported to play important roles in chemical weathering and carbon cycling processes (Dolhi et al., 2012).

Assessments of diversity and distribution of microorganisms in glaciers are fundamental to studies related to global warming. New isolates have been obtained from practically all studied glaciers around the world, and some of them have important physiological and biotechnological characteristics (Garcia-Descalzo et al., 2013).

It has been studied how psychrophiles respond to sudden heat shock or cold shock. Various types of adaptations to temperature rise have been described, both at cellular and community level. At the individual level, the main mechanisms of adaptation to warm temperature include increased synthesis of Heat Shock Proteins (HSPs) (Garcia-Descalzo et al., 2014); pigment production (Martin-Cerezo et al., 2015); and the modification of molecular interactions (Garcia-Descalzo et al., 2011).

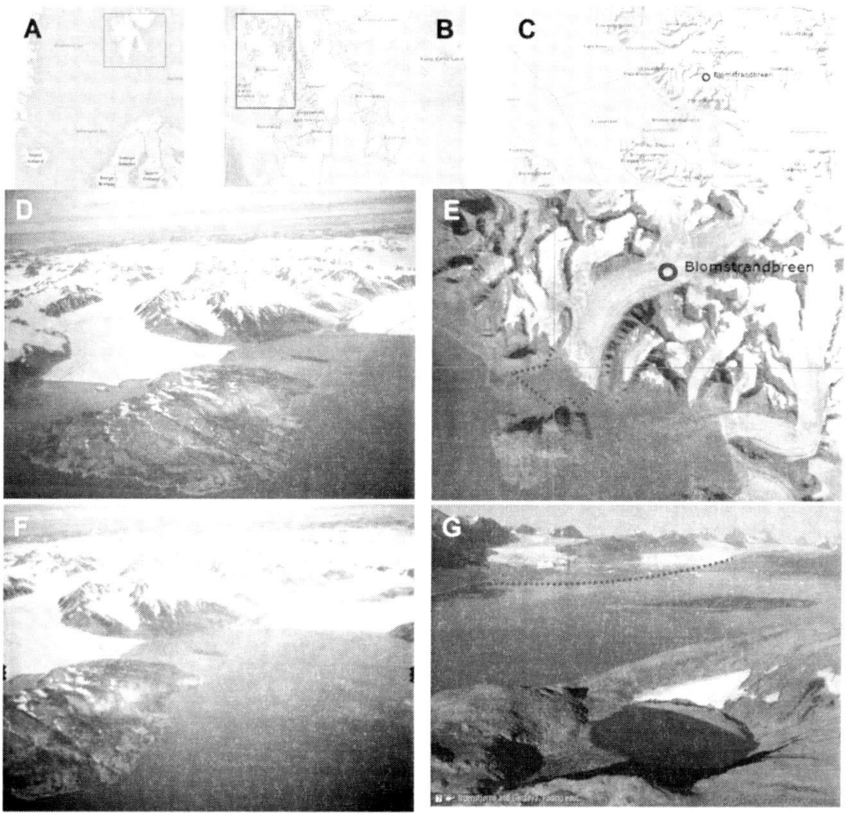

Front positions of glaciers in the year 1936 are marked in red. Photographs belong to the TopoSvalbard archive (http://toposvalbard.npolar.no/).

Figure 1. Retreat of Arctic glaciers. Maps of Svalbard (A, B) and localization of the studied glacier (C). Maps of Blomstrandbreen, Kongsfjord in the Svalbard archipelago showing the retreat of glaciers since the year 1936 (D, F) to 2014 (E, G).

With respect to microbial populations, when temperature rises, it can be observed a change in the three main parameters that ecologically describe a community composition: richness or α-diversity (number of species or operational taxonomic units (OTUs) compared with the number of individuals in the community), abundance (number of individuals -clones- per sample) and diversity (variation of life forms within a given ecosystem). This last parameter, has been measured in our experiments by the Shannon index H´. The maximum Shannon's diversity for a sample is found when all species (OTUs) are equally abundant. Values of the Shannon's diversity index for real communities typically fall between 1.5 and 3.5

4.1. Arctic Glaciers

We have performed several experiments in order to study how microbial communities in a glacier vary depending on the area studied; which in turn represents growth areas of glacier (accumulation area) and areas of regression (ablation zone and glacial lake).

Glacier microorganisms from Blomstrandbreen, Kongsfjord (Figure 1), including both bacteria and small eukaryotes, were identified by sequencing of the 16S rRNA and 18S rRNA respectively; following the methodology described elsewhere (Garcia-Descalzo et al., 2013). Thus, we have observed that the diversity of microorganisms in the area of the glacier snow accumulation (Figure 2A), is greater than in the ablation zone; and this in turn is higher than in the area of liquid water. This result is similar for bacteria (Figure 2A) and eukaryotes (Figure 2B).

4.2. Antarctic Glaciers

There has been increasing interest in characterizing retreating ice fronts of deglaciated forefields (Bradley et al., 2014). With the aim of getting to know how both the richness and the abundance of microorganisms vary in a glacier due to climate change, we have performed several comparisons between glaciated and non glaciated areas during a period of glacier melting, in summer.

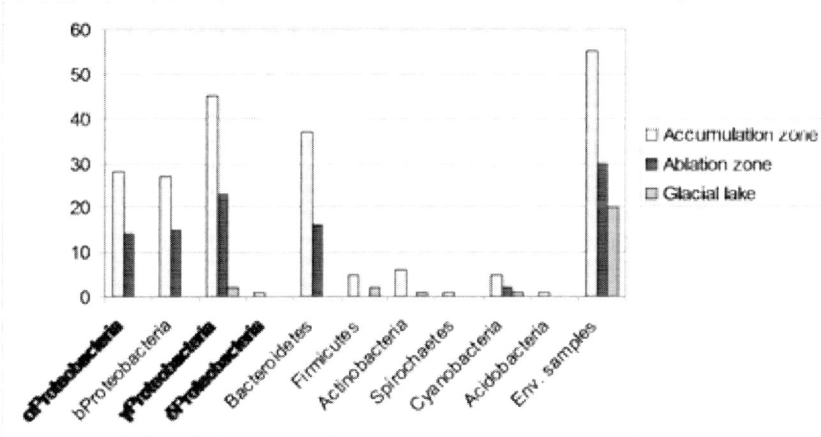

a

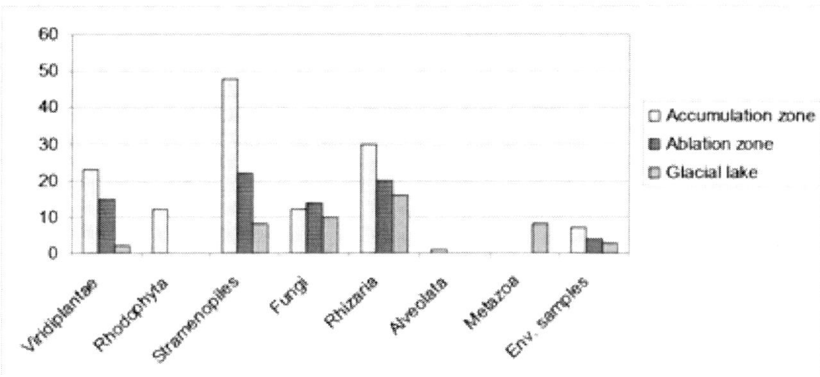

b

Figure 2. Microbial abundance and diversity in glacier Blomstrandbreen, Kongsfjord, Svalbard archipelago. Relative 16S rRNA (A) and 18 rRNA (B) abundance of major bacterial and eukaryotic groups in gene libraries from ice samples.

For our experiments, one of the glaciers located in Livingston Island, South Shetland Archipelago was selected. The chosen site for the study is suitable for these experiments because it allows the research of a transect that crosses the areas of accumulation and ablation of the glacier, and ends in a closed bay, similar to an inland lake.

We compared 10 samples taken from the accumulation zone with 10 samples from the ablation zone, close to the lateral moraine (Figure 3).

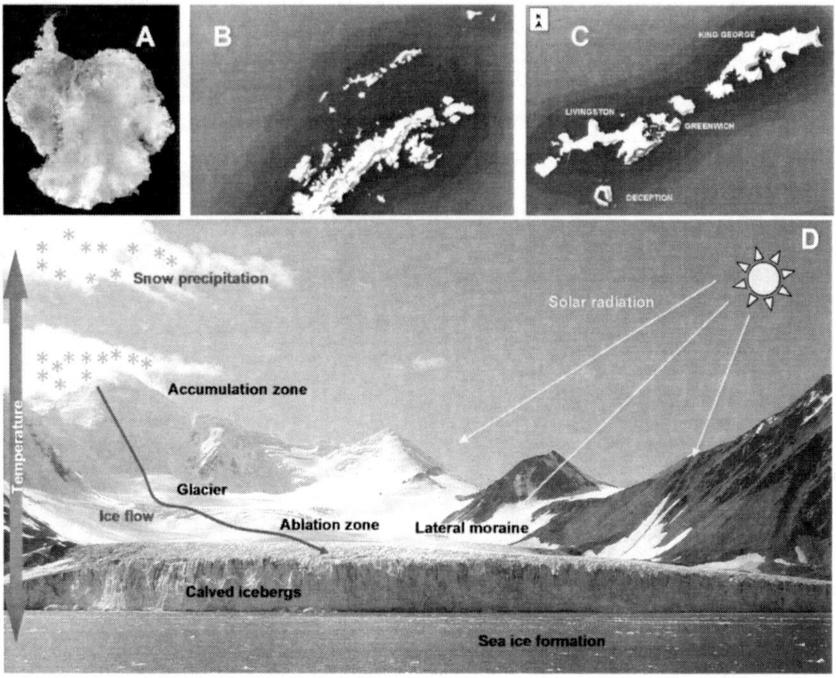

Figure 3. Example of an Antarctic glacier in Livingston Island, South Shetland archipelago. Map showing (A) the Antarctic continent, (B) Antarctic peninsula and (C) Shouth Shetland Archipelago. (D) Schematic picture of the glacier.

Results show that when comparing the 10 samples taken from the accumulation zone versus the 10 samples from the lateral moraine, the differences in the number of bacterial clones is not statistically significant (Figure 4A), while the difference in the number of OTUs is significant (**p = 0.0054) (Figure 4B). The analysis of eukaryotes, revealed that the difference in the number of clones is not statistically significant (Figure 4A), while the difference in the number of OTUs is very significant (***p < 0.0001) (Figure 4B).

4.3. High Mountain Glaciers

Global warming and associated environmental changes are predicted to have strong impacts on high-altitude ecosystems (IPCC, 2007). For instance, European mountains have already experienced an increase in temperature of 1-

2°C during the 20th century (Beniston et al., 1997; Djukic 2012). Our knowledge about the microorganisms living in mountain glaciers is still rudimentary, mostly because of logistical challenges imposed by its inhospitable climate, and the presence of a multi-year ice cap. Comparisons of geographically distinct glaciers worldwide have shown a great variation in microbial biomass and community structure (Xiang et al., 2009).

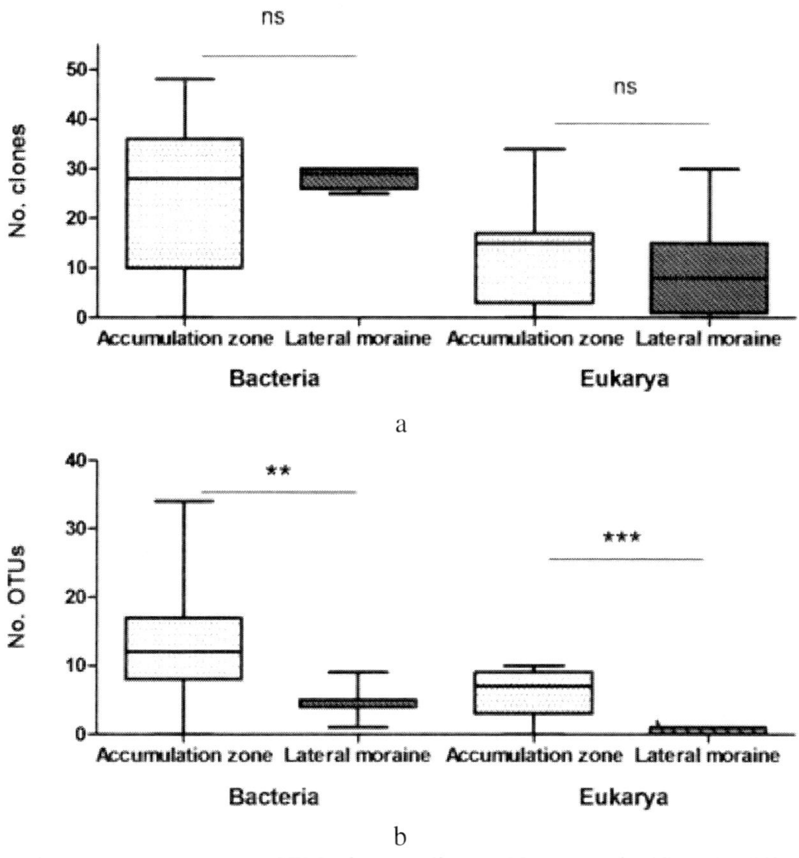

The values represent means ± SEM of ten replicates. Ns = non significantly different; ** $p = 0.0054$; *** $p < 0.0001$ accumulation zone compared to lateral moraine by Student's *t* test.

Figure 4. Statistical comparison of the microbial communities in an Antarctic glacier in Livingston Island, South Shetland archipelago.

We have been performed similar experiments to find the eukaryotic microbial composition of Pyrenean glaciers, comparing glaciers at different altitudes. These experiments have demonstrated that richness, abundance and diversity of microbial communities in higher glaciers differ from the lower glacier communities (Garcia-Descalzo et al., 2013). In general, the snow and ice in these lower glaciers are nearly melted in summer, and they can be a good model to study a deglaciation process. Regardless of the chemical composition of ice, this difference is mainly due to the large number of some groups of characteristic microorganisms, native and dependent on each specific cordillera. There is also a marked decrease in the number of clones, number of OTUs and the Shannon diversity index in lower glaciers. Comparisons of both, the mean number of clones and the mean number of OTUs demonstrated the similarity between the higher glaciers. In general, glaciers located at higher altitude were more diversely populated than lower glaciers.

Conclusion

- Numerous microorganisms live in glacial areas and are specially adapted to these extreme environments, where they maintain key ecological functions.
- Most of environmental microorganisms can not be cultured in laboratory conditions. However, thanks to DNA sequencing methods, their identity is being known.
- Glaciers are inhabited by microorganisms of all domains of life, including representatives of bacteria, archaea, unicellular eukaryotes and viruses.
- Climate change is severely affecting glaciers and their microbial communities. Richness, abundance and diversity of microbial communities are being threatened due to glacier retreat.

Acknowledgments

This research was supported by grants CTM/2008-00304/ANT, CTM 2010-12134-E/ANT and CTM2011-16003-E from the Spanish Ministerio de Economía y Competitividad (MINECO). E.G.L. is recipient of a MINECO

Fellowship (Grant PTAT2010-03424, Programa Nacional de Contratación e Incorporación de RRHH). We are due to Christiane Hübner and Christian Zoelly from the Norwegian Sverdrup Station (Ny Ålesund, Svalbard) for their skillful assistance. We thank also the staff of the Juan Carlos I Research Station (Livingston Island, South Shetland, Antarctica) for their logistic support.

References

Alcazar, A., Garcia-Descalzo, L., Cid, C. 2009. Microbial evolution and adaptation in icy worlds. In: Simon Hegedûs and Jakob Csonka (ed.) Astrobiology: Physical Origin, Biological Evolution and Spatial Distribution. Nova Science Publishers, Inc, pp. 81-95.

Anesio, A.M., Laybourn-Parry, J. 2012. Glaciers and ice sheets as a biome. *Trends in Ecology and Evolution*. 27, 219-225.

Bachy, C., López-García, P., Vereshchaka, A., Moreira, D. 2011. Diversity and vertical distribution of microbial eukaryotes in the snow, sea ice and seawater near the North Pole at the end of the polar night. *Front Microbiol*. 2,106.

Barros, J.A. y Morita, R.Y. 1978. Microbial life at low temperatures: ecological aspects. In: *Microbial life in extreme environments*. Ed. D. J. Kushner. London: Academic Press, pp. 9-71.

Beniston, M., Diaz, H.F., Bradley, R.S. 1997. Climatic change at high elevation sites: an overview. *Clim Change*. 36, 233-251.

Boetius, A., Anesio, A.M., Deming, J.W., Mikucki, J.A., Rapp, J.Z. 2015. Microbial ecology of the cryosphere: sea ice and glacial habitats. *Nature Reviews Microbiology*. 13, 677-90. doi: 10.1038/nrmicro3522.

Bradley, J.A., Singarayer, J.S., Anesio, A.M. 2014. Microbial community dynamics in the forefield of glaciers. *Proc. R. Soc. B* 281: 20140882. http://dx.doi.org/10.1098/rspb.2014.0882.

Branda, E., Turchetti, B., Diolaiuti, G., Pecci, M., Smiraglia, C., Buzzini, P. 2010. Yeast and yeast-like diversity in the southernmost glacier of Europe (Calderone Glacier, Apennines, Italy). *FEMS Microbiol Ecol*. 72, 354-369.

Butinar, L., Spencer-Martins, I. Gunde-Cimerman, N. 2007). Yeasts in high Arctic glaciers: the discovery of a new habitat for eukaryotic microorganisms. *Antonie van Leeuwenhoek*. 91, 277-289.

Cameron, K.A., Hodson, A.J., Osborn, A.M. 2012. Structure and diversity of bacterial, eukaryotic, and archaeal communities in glacial cryoconite holes from the Arctic and the Antarctic. *FEMS Microbiology Ecology* 82, 254-267.

de García, V., Brizzio, S., Libkind, D., Buzzini, P., van Broock, M. 2007. Biodiversity of cold-adapted yeasts from glacial meltwater rivers in Patagonia (Argentina). *FEMS Microbiol Ecol.* 59, 331-341.

De Maayer, P., Anderson, D., Cary, C., Cowan, D.A. 2014.Some like it cold: understanding the survival strategies of psychrophiles. *EMBO Rep.* 15, 508-17.

Djukic, I., Zehetner, F., Watzinger, A., Horacek, M., Gerzabek, M.H. 2012. *In situ* carbon turnover dynamics and the role of soil microorganisms therein: a climate warming study in an Alpine ecosystem. *FEMS Microbiol Ecol.* doi: 10.1111/j.1574-6941.2012.01449.x.

Dolhi, J.M., Ketchum, N., Morgan-Kiss, R.M. 2012. Establishment of microbial eukaryotic enrichment cultures from a chemically stratified Antarctic lake and assessment of carbon fixation potential. *J Vis Exp.* 62.

Edwards, A., Anesio, A.M., Rassner, S.M., Sattler, B., Hubbard, B., Perkins, W.T., Griffith, G.W. 2011. Possible interactions between bacterial diversity, microbial activity and supraglacial hydrology of cryoconite holes in Svalbard. *The ISME journal.* 5,150-160.

Foght, J.M., Aislabie, J., Turner, S., Brown, C.E., Ryburn, J., Saul, D.J., Lawson, W. 2004. Culturable bacteria in subglacial sediments and ice from two southern hemisphere glaciers. *Microb Ecol.* 47, 329-340.

García-Descalzo, L., Alcazar, A., Baquero, F., Cid, C. 2011. Identification of *in vivo* HSP90-interacting proteins reveals modularity of HSP90 complexes is dependent on the environment in psychrophilic bacteria. *Cell Stress Chaperones.* 16,203-218.

García-Descalzo, L., García-López, E., Postigo, M., Baquero, F., Alcazar, A., Cid, C. 2013 Eukariotic microorganisms in cold enviroments: examples from Pyrenean glaciers. *Frontiers in Microbiology.* 4, 55. doi: 10.3389/fmicb.2013.00055.

García-Descalzo, L., Alcazar, A., Baquero, F., Cid, C. Biotechnological applications of cold-adapted bacteria. Extremophiles: Sustainable Resources and Biotechnological Implications, First Edition. Edited by Om V. Singh. 2013 Wiley-Blackwell. Published by John Wiley and Sons, Inc. pp. 159-174.

García-Descalzo, L., García-López, E., Alcazar, A., Baquero, F., Cid, C. 2014. Proteomic analysis of the adaptation to warming in the Antarctic bacteria

Shewanella frigidimarina. *BBA Proteins and Proteomics*. 1844, 2229-2240.

Gilichinsky, D., Wagene, S. 2006. Microbial life in permafrost: A historical review. *Permafrost and periglacial processes*. 6, 243-250.

Hambrey, M., Alean, J. 2004. Glaciers. Cambridge: Cambridge University Press.

IPCC, 2007: Climate change 2007: the physical science basis. Contribution of Working Group I to the Fourth Assessment Report of the Intergovernmental Panel on Climate Change (Solomon, S., Qin, D., Manning, M., Chen, Z., Marquis, M., Averyt, K.B., Tignor, M. and Miller, H.L. (eds.). Cambridge University Press, Cambridge, UK and New York.

IPCC, 2014: Climate Change 2014: Synthesis Report. Contribution of Working Groups I, II and III to the Fifth Assessment Report of the Intergovernmental Panel on Climate Change. Core Writing Team, R.K. Pachauri and L.A. Meyer (eds.). IPCC, Geneva, Switzerland, 151 pp. http://www.ipcc.ch.

Immerzeel, W.W., van Beek, L.P.H, Bierkens, M.F.P. 2010. Climate change will affect the Asian water towers. *Science*. 328, 1382. doi: 10.1126/science.1183188.

Lefranc, M., Thenot, A., Lepere, C., Debroas, D. 2005. Genetic diversity of small eukaryotes in lakes differing by their trophic status. *Applied and environmental microbiology*. 71, 5935-5942.

Ma, L., Rogers, S. O., Catranis, C. M., Starmer, W. T. 2000. Detection and characterization of ancient fungi entrapped in glacial ice. *Mycologia*. 92, 286-295.

Maccario, L., Sanguino L., Vogel, T.M., Larose, C. 2015. Snow and ice ecosystems: not so extreme. *Research in Microbiology*. 166, 782-795.

Mader, H.M., Pettitt, M.E., Wadham, J.L., Wolff, E.W., Parkes, R.J. 2006. Subsurface ice as a microbial habitat. *Geology*. 34,169-172.

Margesin, R., Fonteyne, P.A., Schinner, F., Sampaio, J.P. 2007. Novel psychrophilic basidiomycetous yeasts from Alpine environments: *Rhodotorula psychrophila* sp. nov., *Rhodotorula psychrophenolica* sp. nov. and *Rhodotorula glacialis* sp. nov. *Int J Syst Evol Micr*. 57, 2179-2184.

Martín-Cerezo, M.L., García-López, E., Cid, C. 2015. Isolation and Identification of a Red Pigment from the Antarctic Bacterium *Shewanella frigidimarina*. *Protein Pept Lett*. 22, 1076-82.

Medinger, R., Nolte, V., Pandey, R.V., Jost, S., Ottenwälder, B., Schlötterer, C., Boenigk, J. 2010. Diversity in a hidden world: potential and limitation of next-generation sequencing for surveys of molecular diversity of eukaryotic microorganisms. *Mol Ecol.* 19, 32-40.

Michelsen, A., Rinnan, R., Jonasson. S. 2012. Two Decades of Experimental Manipulations of Heaths and Forest Understory in the Subarctic Anders. *AMBIO*. 41, 218-230. doi: 10.1007/s13280-012-0303-4.

Miteva, V.I., Sheridan, P.P, Brenchley, J.E. 2004. Phylogenetic and physiological diversity of microorganisms isolated from a deep Greenland glacier ice core. *Appl. Environ. Microbiol.* 70,202-213.

Miteva. 2008. Bacteria in Snow and Glacier Ice. Psychrophiles: from Biodiversity to Biotechnology. Springer-Verlag Berlin Heidelberg.

Morita RY. 1975. Psychrophilic bacteria. *Bacteriol Rev*. 39, 144-167.

Price, P.B. 2000. A habitat for psychrophiles in deep Antarctic ice. *Proc Natl Acad Sci US* 97, 1247-1251.

Russell, N.J. 1990. Cold adaptation of microorganisms. *Phil. Trans. R. Soc. Lond.* 326, 595-611.

Russell, N.J. 1992. Physiology and mollecular biology of psycrophilic micro-organisms. In: *Mollecular biology and biotechnology of extremophiles.* Champman and Hall, New York. pp. 203-221.

Russell, N.J. 1993. Biochemical differences between psychrophilic and psychrotolerant microorganisms. In: *Trends in Microbiology Ecology*. Ed. Gerrero R., Pedros-Alio C. Madrid: Spanish Society for Microbiology. pp. 29-32.

Russell, N.J., Hamamoto T. 1998. Psycrophiles. In: *Extermophiles: Microbial Life in Extreme Environments*. Ed. Horikoshi K. and Grant W. D. Wiley-Liss. pp. 25-45.

Schmidt-Nielse, S. 1902. *Zentralbl. Bakteriol. Parasitenkd. Infektionskr. Hyg.* 9, 145.

Skidmore, M., Foght, J. M., Sharp, M. J. 2000. Microbial life beneath a high arctic glacier. *Appl Environ Microb.* 66, 3214-3220.

Stibal, M., Sabacka, M., Kastovska, K. 2006. Microbial communities on glacier surfaces in Svalbard: impact of physical and chemical properties on abundance and structure of cyanobacteria and algae. *Microb. Ecol.* 52, 644-654.

Taton, A., S. Grubisic, E. Brambilla, R. De Wit, A. Wilmotte. 2003. Cyanobacterial diversity in natural and artificial microbial mats of Lake Fryxell (McMurdo Dry Valleys, Antarctica): a morphological and molecular approach. *Appl. Environ. Microbiol.* 69, 5157-5169.

TopoSvalbard. http://toposvalbard.npolar.no/.

Turchetti, B., Buzzini, P., Goretti, M., Branda, E., Diolaiuti, G., D'Agata, C., Smiraglia, C., Vaughan-Martini, A. 2008. Psychrophilic yeasts in glacial environments of Alpine glaciers. *FEMS Microbiol Ecol*. 63, 73-83.

Wagner, A. O., Hofmann, K., Prem, E., Illmer, P. 2012. Methanogenic activities in alpine soils. *Folia Microbiol*. 57, 371-373.

Wynn-Williams DW. 1990. Ecological aspects of Antarctic microbiology. *Adv. Microbial Ecol*. 11, 71-146.

Xiang, S., Shang, T., Chen, Y., Jing, Z., Yao, T. 2009. Dominant Bacteria and Biomass in the Kuytun 51 Glacier. *Appl Environ Microb*. 75, 7287-7290. doi:10.1128/AEM.00915-09.

Zhang, X., Ma, X., Wang, N., Yao, T. 2009. New subgroups of Bacteroidetes and diverse microorganisms in Tibetan plateau glacial ice provide a biological record of environmental conditions. *FEMS Microbiol. Ecol*. 67, 21-29.

In: Glaciers
Editor: Nicholas Doyle
ISBN: 978-1-63484-941-8

Chapter 5

MOUNTAIN GLACIERS OF THE SE ALTAI, RUSSIA: THE HOLOCENE CHRONOLOGY AND EFFECTS ON THE TOPOGRAPHY CHANGES

***R. K. Nepop*[1,2,*] *and A. R. Agatova*[1,2]**
[1]Institute of Geology and Mineralogy, Novosibirsk, Russia
[2]Ural Federal University, Yekaterinburg, Russia

ABSTRACT

Being potentially good climatic indicators, mountain glaciers give the opportunity to analyze climatic variations which controlled their dynamic. This paper presents the results of investigations of mountain glaciers within the southeastern part of Russian Altai (SE Altai). This is one of the most northern centers of modern glaciations in Central Asia, where evidences of significantly larger Holocene mountain valley glaciers are clearly preserved in topography as well as traces of more extensive semi- ice-sheet and valley-piedmont Pleistocene glaciations. We present the radiocarbon chronology of the Holocene glaciers dynamic in the region. Applying numerical simulation, we also evaluate the influence of modern mountain valley glaciers on topography changes. Obtained values indicate the possible low estimates of more significant

* Koptyuga av., 3, Novosibirsk, Russia, 630090. agatr@mail.ru.

relief changing activity of more powerful Holocene and Pleistocene glaciers within the SE Altai.

Summarizing the results of detailed geomorphological investigations and analysis of expanded radiocarbon dataset of moraines and position of the upper timber limit within the SE Altai we argue for presence of three Holocene glacial stages when glaciers were larger than modern ones. Obtained data was also correlated with the periods of soils and peats formation as well as intensifications of slope processes. The resulted chronology of glacier fluctuations refutes the traditional concept of the Russian Altai Holocene glaciations as a consecutive retreat of the late Würm (Sartan according to the local stratigraphic scheme) glaciers and argues their complete degradation at the head of trough valleys by the beginning of the Holocene. Moraine complexes of three Holocene glacial stages are morphologically expressed in topography of the SE Altai. They correlate with three identified periods of glacial advances in the second half of the Holocene: from 4900 to 4200 cal. years BP (Akkem stage), from 2300 to 1700 cal. years BP (Historical stage) and in the 13th – 19th centuries (Little Ice Age (LIA) or Aktru stage). Repeated forest regrowth in the presently glaciated area indicates significant retreat or even complete glacier degradation during interstage warming.

Numerical simulation of the ice movement in the mountain-valley glacier and its interaction with the glacier bed allows estimating the glacial erosion rate which characterizes the contribution of glaciers to the topography denudation in mountain provinces affected by glaciations. One-dimensional numerical model of idealized glacier describes the ice flow along central flow line. Physical constants, hypsometric data and other parameters of model glacier are closely corresponded to that ones of the real SE Altai modern glaciers as well as to climatic, geological and geomorphological conditions of this region. Absolute glacier erosion rate due to numerical simulation is about $1.0 \cdot 10^{-3}$ $m \cdot a^{-1}$ with the peak value $2.0 \cdot 10^{-3}$ $m \cdot a^{-1}$ near the equilibrium line. The reaction time to a stepwise change of the glacier mass balance was calculated as about 10^2 years. Applying this model, we have also qualitatively analyzed the evolution of the glacial-valley longitudinal profile.

Obtained numerical estimates were compared to available regional data which allowed outlining the range of possible values of regional glacial erosion rates. These estimates also correspond with the results obtained for other glaciated areas of Central-Asia.

Keywords: mountain glaciations, glacier fluctuations, climatic variations, upper timber limit, radiocarbon chronology, numerical simulation, erosion rate, Holocene, Russian Altai

INTRODUCTION

Collection and analysis of information about the Holocene dynamics of glaciations and its subsequent synthesis in form of paleogeographic and paleoclimatic reconstructions are relevant not only for understanding the past climate changes but also for their prediction in future. In contrast to a hypothesis of a stable Holocene climate, as indicated by oxygen isotope records from the Greenland Ice core (Johnsen et al., 1997), a growing number of studies (e.g., Mayewski et al., 2004; Solomina et al., 2015 and references therein) have demonstrated that distinct periods of climate change occurred repeatedly throughout the Holocene. Mountain glaciers are potentially good climatic indicators, but most of the information on the dynamic responses of mountain glaciers to past climate changes comes from European glaciers which only form 3% of the world's glaciated area (Dyurgerov and Meier, 1997). To obtain an integral picture of spatial variability of the Holocene climate within inner Eurasia it is necessary to take full advantage from glacial dynamic studies in all segments of the Central Asia Orogenic Belt. In this context Russian Altai is still poorly investigated despite the fact that it is one of the centers of modern glaciations in Central Asia and it was affected by powerful glaciations in the past. This paper seeks to expand the current Holocene climatic records worldwide by presenting data from the SE Altai, Russia.

Modern glaciation area within Russian Altai is significantly less than an area of the Holocene glaciers in times of their maximal expansion and several orders of magnitude less than an area affected by the Pleistocene glaciations (Lehmkuhl et al., 2007; Okishev, 2011; Agatova et al., 2012; Solomina et al., 2015 and others). Nevertheless even today glacial activity is one the major relief-changing factor in this high mountain province. Studying modern glaciers also forms the basis for understanding Quaternary glacial landforms.

The first and so far the unique for Russian Altai qualitative determinations of glacial erosion rate were done in the second half of the last century (Dushkin, 1974). These researches were based on measurements of quantitative content of moraine material in the ice of glacier's tongue in its active terminal parts. At the same time the development of plastic ice flow theory (Nye, 1969), experiments on determining the physical parameters of moving ice, field observations of mountain-valley glaciers as well as rapid progress in computer technology make it possible utilizing numerical modeling in glaciological investigations. Numerical simulations are increasingly used to model the glacier mass balance (Greuell, 1992; Caidong and Sorteberg, 2010; and others), determine its dynamic characteristics (Albrecht et al., 2000; De Smedt and Pattyn, 2003; and others), front position (Oerlemans, 1997; MacGregor et al., 2000; and others), and in many other cases. A review of various mathematical models and numerical methods applied in glaciology is given for example in (Paterson, 1994).

Numerical simulation of the ice movement in the mountain-valley glacier and its interaction with the glacier bed, presented in this paper, allows *i)* calculate the contemporary glacial erosion rate for the SE Altai; *ii)* estimate the characteristic time of their response to climate change; and *iii)* qualitatively asses the evolution of the longitudinal profile of a glacial valley in the time interval about 10^3-10^5 years, using the parameters of modern glaciation of the SE Altai. Obtained numerical estimates allowed establishing the glaciers contribution into the topography denudation and development of longitudinal profile of mountain valleys in glacial areas of the SE Altai.

Applied Methods and Techniques

Geomorphological methods were used for reconstruction of glacier advances for different stages. The distribution of moraines and proglacial landforms in the North Chuya range was investigated based on interpretation of aerial photographs, topographic maps at a scale of 1:25000, and field investigations which included mapping and sedimentological analysis of terminal and flank moraines in different states of preservation and vegetation cover. Moraines in each glacial valley were initially classified using their elevation, geographical position, the degree of weathering of moraine material, composition and shape of moraine component parts, and ecological characteristics (vegetation cover and lichenometry). Genesis determination of other landforms composed of unconsolidated fragmental rock deposits, such as rock glaciers, landslides, debris flows, etc., was also made.

Lichenometry was used to separate the moraine complexes of the Historical and Aktru stages, based on their relative age. The *Rizocarpon geographicum*, regarded as one of dominant species of crustose lichen in the Altai (Solomina, 1988), was identified from the shapes of the areola and aphothecium with a hand lens. The long and short axes were measured using a Vernier micrometer (0.05 mm increments). The mean values of the five largest long axes were used as lichen data.

Lichen analysis of *Rhizocarpon geographicum* confirmed the different ages of moraine complexes associated with the Aktru and Historical stages in some valleys of the North Chuya range where the youngest "Aktru" moraines partly cover the "Historical" ones. The size of *Rhizocarpon geographicum* lichen at the boulders on the surface of "Historical" moraines is significantly larger than those on the surface of younger "Aktru" moraines (Agatova et al., 2012).

Radiocarbon ages of wooden fragments were used for determining the periods favourable for growing forest vegetation in the heads of troughs. These periods limit the times of climate deterioration and possible glacier advances. All dates were measured at the Cenozoic Geochronology Center SB RAS, Novosibirsk, and at the Institute of Geography RAS, Moscow, by the conventional ^{14}C method. Determination of carbon residual activity in both laboratories was done with the QUANTULUS-1220. This paper presents both calibrated (cal BP) and uncalibrated radiocarbon ages for all samples. A ^{14}C half-life of 5570 years was used for age calculations and for converting the half-life to 5730 years a coefficient of 1.03 should be taken. Ages are calculated relative to the year AD1950. The conventional radiocarbon ages were calibrated applying the CALIB Rev 5.0 program (Stuiver and Reimer, 1993), with the intcal09 calibration dataset (Reimer et al., 2009). The corresponding lowest and highest limits of the 2-sigma standard deviation of calibrated ages are reported and using the same program we have also calibrated all the previously available radiocarbon dates.

For specifying chronological limits of time periods favourable for growing forest vegetation in the heads of troughs life span of dead trees in modern glacial zone was estimated: tree rings from collected wooden fragments were counted when it was possible.

Numerical simulation of glacier ice movement and its interaction with the valley bottom was applied for estimating the glacial erosion rate. This simulation and data processing were realized in an integrated development environment software products «Delphi7» (Nepop, 2007).

STUDY AREA

Orography and Climate

The Altai Mountains are the northern part of the Central Asia collision belt and geographically belongs to the mountains of Southern Siberia. They stretch northwest more than 1500 km across the borders of Mongolia, China, Kazakhstan and Russia, and form a wedge shape narrowest in the southeast and widest in the northwest (Figure 1). The elevation increases in the opposite direction from 400 m a.s.l. to 4000 m a.s.l. Modern climate of Altai Mountains is determined by its intracontinental position, with the main moisture transfer from the west (Atlantic Ocean), and to a lesser degree from the north (Arctic), with a dominant influence of the Mongolian anticyclone giving rise to increasing aridity southeastwards and complicated latitude-longitude orographic climatic zoning.

Figure 1. Altai Mountains location.

The high-mountain southeastern part of the Russian Altai includes the Chuya and the Kurai intermountain depressions surrounded by ridges with altitudes about 3500-4200 m a.s.l. (Figure 2). The Belukha peak (4506 m a.s.l.) is the highest mountain in the Asian part of Russia. Deeply incised valleys represent a system of imbedded troughs including up to 3 generations within the South Chuya range (Agatova, 2005). The floor of the Chuya intermountain depression - the largest in the Altai Mountains - is located at 1750-2000 m a.s.l. lowering westwards where the incision of the Chuya river-bed is about 1735 m a.s.l. The altitude of the Kurai depression is about 1500-1600 m a.s.l. These intermountain depressions are separated by the Chagan-Uzun massif with the highest Sukor summit (2920 m a.s.l.) in its northern part.

In the western part of the SE Altai the stony steppe on the floor of the Kurai depression grades into taiga vegetation on the ranges slopes. In the western mountain frame of the Chuya depression forests have an insular distribution, vanishing completely in the southeastern part. The plateau-toppled highlands are covered by alpine meadows. The higher relief layer is represented by alpine landscapes with mountain tundra vegetation changing with height into glacial zone.

Figure 2. Sketch map of the SE Altai showing the locations of study sites discussed in the paper. A- Aktru valley; B- Boguty valley; C- mouth of the Chikty river; D- Berel valley; E- Kuraika valley; H- Shavla valley; I- Chuya Valley, mouth of the Kuektanar river; J- the source of Chuya river and the valleys of its tributaries in the eastern part of the Chuya depression; K- Kurkurek, Jan-Karasu valleys and Ehtykkel stow; L- Tara valley; M- Maashey valley; N- Aksai valley; P- foot of the Kurai range; Q- Valleys of Akkem and Kucherla rivers; R- Mensu valley; S- Sofiyskiy glacier; T- Taldura valley; U- Ulandryk valley.

In spite of the arid climate (the mean annual precipitation is less than 200 mm in the floor of intermountain depressions), the high altitudes of the ridges are favorable to glacier formation. Modern glaciers feed the tributaries of the Chuya, Argut and Katun rivers which basins include most of rivers of the SE Altai.

The ice-sheet mid-Pleistocene and the valley-piedmont late-Pleistocene glaciations greatly affected the SE Altai landscapes (Devyatkin, 1965). Smaller Holocene glaciers mainly occupied the heads of troughs. In some glacial valleys up to two or three moraine complexes regarded as the youngest Holocene stages can be clearly identified. Moraines of the Aktru stage, nearest to the modern glaciers terminus are clearly morphologically expressed in all glacial valleys and often partly cover moraines of the Historical stage.

Characteristics of the SE Altai Contemporary Glaciation

About 75% of the modern glaciated area in the Russian Altai, with a total surface 910 km² (Revjakin et al., 1979), is concentrated in its highest southeastern part. Glaciers are distributed here in an irregular manner. The position of the SE Altai within the border area affected by Mongolian anticyclone and by humid air masses coming from Atlantic Ocean leads to clearly expressed longitudinal climatic zoning. The mean annual precipitation near the snow line decreases along the W-E axis from 2000 mm to less than 500 mm (Narozny, Osipov, 1999), and as a result, the altitude of the snow line rises from 2400 m a.s.l. in the western part of Katun ridge to 3350 m a.s.l. on the Mongolian border. Accordingly, in the same direction an area and volume of glaciers is reduced. The similar distribution of modern, Holocene, and late Pleistocene glaciers argue for the same oroclimatic zoning and moisture deficit in this area from the beginning of the late Pleistocene (Agatova, 2005).

Regional climate peculiarity is that winters here are too cold and dry to significantly affect the glaciers mass balance. Thus the SE Altai glaciers can be characterized as summer accumulation-type glaciers due to the coincidence of accumulation and ablation seasons in summer (Galahov and Muhametov, 1999; Pattyn et al., 2003). Current knowledge about the behavior of these glaciers is limited - they are reported to be more vulnerable to the current global warming trend than the better-known (maritime) winter-accumulation-type glaciers (Fujita and Ageta, 2000).

Generally the SE Altai is characterized by more arid climate (the mean annual precipitation is about 150-200 mm within the floor of Kurai and Chuya intermountain depressions and up to 1200-1500 mm in the crest areas of surrounding ridges (Rusanov, 1961)) in comparison to the NE Altai, where on the most part the mean annual precipitation is about 1000 mm. Under conditions of moisture deficit, the main factors controlling glacier cover are:

1) Absolute altitudes of the ridges. Thus the crest area of the South Chuya range with the altitudes 3000-3560 m a.s.l. is one of the centres of modern glaciation, whereas in the neighboring Chagan-Uzun massif (2700-2920 m a.s.l.) there are no contemporary glaciers;
2) Slope insolation which is particularly important in the case of the sub-latitudinal trending of the most mountain ranges within the SE Altai;
3) Asymmetric structure of mountain ranges - gentle northern slopes are much longer than steep southern ones, which causes a larger surface

for the glacier basins with northern exposure, as well as concentration of solid precipitation on the northern leeward slopes.

Thus, the overwhelming majority of the glaciers are located on the northern slopes of the highest ranges of the SE Altai: Katun, South Chuya and North Chuya. All types of glaciers are presented here – from large mountain-valley glaciers and glaciers of flattened mountains tops to small cirque and hanging ones. The cirque and hanging glaciers are dominated by the number, when mountain-valley ones - by the area. The total ice volume on the northern slopes of the North Chuya range is twice as much as on the southern ones. Within the South Chuya range located to the east this ratio is about 7:1 (Nikitin et al., 2000). The thickness of the glaciers is relatively small. Radio-echo sounding revealed the maximal ice thickness not more than 100-150 m in the most of mountain-valley glaciers, about 50-70 m in cirque glaciers, and 25-50 m in small hanging glaciers (Catalogue of glaciers). Radio-echo sounding also detected mainly one – maximum two layers with different ice temperature and water saturation in glaciers of the SE Altai. For example in Maly Aktru glacier, North-Chuya range, the upper bounder of the water saturated layer is located about 20-40 m beneath the ice surface with its total thickness about 160 m (Nikitin and Tatarinov, 1982).

Generally the presence of such layers complicates numerical simulation (Huybrechts et al., 1989), but simple structure of regional glaciers allowed developing in this study an ordinary single-layer numerical model.

History of Investigations

Holocene glacier variations in the Russian Altai have been investigated and debated by many researchers (Mjagkov, 1936; Tjumencev, 1936; Vardanyants, 1938; Dushkin, 1965; Ivanovsky, 1967, 1993; Ivanovsky and Panychev, 1978; Ivanovsky et al., 1982; Okishev, 1982; Butvilovsky, 1993; Solomina, 1999; Galahov et al., 2005; 2008 and many others). However, until the beginning of this century due to the small number of absolute dates the Holocene chronology of glaciers dynamic within Russian Altai was reconstructed mainly on the basis of terminal moraines relative position in comparison with the modern glaciers, absolute altitudes of these moraine complexes, their state of preservation and forest coverage, as well as calculations of the equilibrium line altitude (ELA) depression.

The Holocene glacial stages in the Russian Altai are traditionally marked out according to the scheme of Würm glacier degradation, suggested by Vardanyants (1938) (Table 1). He identified eight stages of glacial retreat in the Caucasus and correlated them with the Alpine ones, and further extended them to the Russian Altai using data from Tjumencev (1936) and Mjagkov (1936), who studied the glaciations of the Katun range as part of the USSR Geological Survey (1929-1932) and a USSR Academy of Sciences glacial expedition (1933). Stages in this scheme are based on names of local lakes and rivers.

Table 1. The climatically forced Holocene glacial dynamics and ELA depression in the southeastern part of the Russian Altai after different authors

Glacial stages after (Vardanyants, 1938)	Altitude of terminal moraines of different stages for the North Chuya range. Glacial stages chronology for some Central Asia Mountains after (Shnitnikov, 1957)		Glaciers advances and ELA depressions for Aktru basin, the North Chuya range, after (Ivanovsky, 1967)		Stadial terminal moraines for 2nd megastadial and ELA depressions for the Right Aktru glacier, North Chuya range, after (Okishev, 1982)	
	altitude m a.s.l.	Stage's age (interstage difference ~1800 years)	phase of glaciers advances	ELA depression m	Moraine's order (from the glacier)	ELA depression m
Aktru	2200	17th – 19th	VII	50-70	1	70
Historical	2300	~AD 0	VI	100	2	175
Akkem	1900-2000	~BC 1900	V	250	3	285
Kochurlinskaya	1700	BC 3700-3800	IV	450-470	4	355
Multinskaya	≈1550	BC 5600-5800	III	600	5	430
Ognevskaya	1350	BC 7400-7600	II	650	6	485
First	1250	BC 9200-9400	I	< 700	7	540
Würm maximum	≈1100	BC 11000-11300	-		8	625

Later, Shnitnikov (1957) analyzed the periods of marine transgressions and regressions at the coast of Fennoscandia and suggested a hypothesis of rhythmic general moistening of the continents in the Northern Hemisphere with a period of about 1850 years. The simultaneous humidity fluctuations of continents in the northern hemisphere are expressed by synchronous changes in the levels of seas, lakes and the great oceans; changes of moisture of marshes and peatbogs; and by glacier advances/retreats. He tries to correlate this eightfold 1850-year rhythm of humidity fluctuation to inner Eurasia, including glaciated mountain areas and the scheme of Vardanyants is used for confirmation of this hypothesis. But the eight "expected" stages of glacier degradation cannot be identified in any of the valleys of the Katun range, and the number of terminal moraines should not be used to estimate the absolute ages of these deposits. Nevertheless, some papers (Shnitnikov, 1957; Okishev, 1982) use such a simplified method of dating for estimating the moraine ages with a difference of 1800-1900 years between subsequent deposits.

Due to a lack and often complete absence of absolute dates, relative ages of the moraines have been based on morphological criteria, the moraine's absolute altitudes, and estimations of ELA depression. ELA depression for the same stage within the SE Altai, however, varies between different authors by tens of meters (see Table 1). Obviously such variation excludes the use of this parameter for absolute dating and the correlation of moraines in different valleys. Despite its weaknesses, the Vardanyants scheme had being used for a long time as a basis for paleoglaciological reconstructions within the SE Altai.

Since the 1970s, methods of absolute dating suitable for glacial deposits have become available. The radiocarbon method (Ivanovsky and Panychev, 1978; Ivanovsky et al., 1982; Mikhailov, 1987), dendrochronology (Dushkin, 1965; Adamenko and Syubaev, 1977; Adamenko and Selishchev, 1984) and lichenometry (Solomina, 1988) have been used for studying glacial relief and sediments. Initially, radiocarbon ages for the moraines closest to the modern glaciers in Aktru, Korumdu, Jan-Karasu (North Chuya range), and Berel (Katun range) valleys, and sediments of a moraine-dammed lake in Chikty valley (South Chuya range) were obtained. Subsequently, radiocarbon dates were determined for dead trees toppled by moraines, from rock glaciers and proglacial forefields in glacial valleys of the North Chuya and South Chuya ranges, for peat layers from sediments at foothills of the North and South Chuya ranges and the Kurai intermountain depression, and for fossil soil buried in the Chuya intermountain depression.

The most comprehensive overview of the Holocene climate and environmental changes in the Russian Altai is a monograph (Butvilovsky,

1993). However the author did not directly study moraine complexes in trough valleys of the SE Altai. Using a broad range of analytical techniques for glaciological reconstructions, the interpretation was based mainly on radiocarbon ages of fossil soils and peat within the intermountain depressions and foots of framing ridges. In general, the author attributes the Holocene thermal maximum to the early Holocene and in the Late Holocene he theorizes several glacial stages connected with climate deterioration (about 6000(?), 4200, 500-100 radiocarbon years BP) and believes relatively short duration of periods of intensive cooling and wetting (about 300-500 years) divided by prolonged periods of considerable warming occurred. This hypothesis is mainly supported by our researches. By the end of the 20th century, however, the total number of absolute ages obtained was still insufficient for making reliable paleoglaciological reconstructions, only the age of the youngest Aktru stage (LIA) was reasonably constrained. In the last few years research into Holocene glacial dynamics and climate fluctuations, including absolute dating of moraines and the paleo upper tree limit in trough valleys within the SE Altai, has been undertaken (Galahov et al., 2005, 2008; Nazarov and Agatova, 2008). In addition to direct dating of moraines, attempts to study climatic change in the Russian Altai have been made on the basis of palynological analysis (Schlütz and Lehmkuhl, 2007; Blyakharchuk et al., 2008).

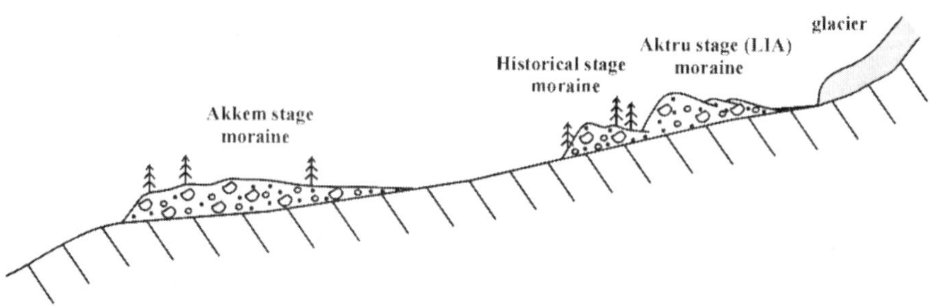

Figure 3. Three stages of the Holocene glacier advances within the North Chuya range, SE Altai expressed in topography after (Agatova et al., 2012).

Up to two or three moraine complexes regarded to the youngest Holocene stages come to confident morphological identification in some mountain valleys within the SE Altai. Moraines of Aktru (LIA) and Historical stages are located in the heads of valleys. Often Aktru moraines partly cover the Historical moraines. In the North Chuya and Katun ranges moraines of the Akkem stage are located several km downstream but are not identifiable in all valleys (Figure 3). In some glacial valleys of the SE Altai the Akkem moraines are not clearly identified in topography. This fact and absence of direct absolute age determinations of Akkem moraines are the main reasons of ongoing discussions about correspondence of distant moraine complexes to the Holocene glaciations as well as about magnitude of glaciers advances during the Akkem stage in different mountain ranges of the SE Altai. Moraines of Aktru stage, nearest to the modern glaciers terminus are morphologically expressed clearly in all glacial valleys of the region.

CHRONOLOGY OF THE HOLOCENE GLACIER DYNAMICS IN THE SE ALTAI

During the last decades, a unique opportunity for investigation of the Holocene glacial history occurred. Ongoing glacier retreat and ice melting in moraines led to extended forefields where meltwater from the glacier terminus remobilizes large amounts of glacial and fluvioglacial sediment. It coursed the exhumation of buried wood macroremnants, allowing the possibility of radiocarbon dating.

Generally among all collected samples we identify the buried tree fragments (including those located *in situ*) and surface ones. Pieces of "*buried*" wood in all valleys are imbricated among boulders on the surface of glaciofluvial deposits or embedded within it. Sometimes samples have melted out directly from the front of the glacier's tongue or been pressed onto the substratum near the ice front. These samples, usually log fragments, show traces of glacier processing: abrasion, twists, ruptures, incisions etc. They are heavily deformed and often covered by a specific crust of peeling wood. "*Surface*" samples were collected from dead trees at the moraine surfaces or at the upper timber limit on trough slopes. These trees were not directly affected by glaciers but were killed by strong deterioration of local climate.

The extended dataset of recently obtained radiocarbon dates together with the previously published ones describes the Holocene glacier dynamics forced by climate changes within the SE Altai (Agatova et al., 2012; Agatova et al., 2014b). Applying dendrochronological analysis with its annual resolution allows specifying the upper timber limit fluctuations and LIA glacier oscillations (Nazarov and Myglan, 2012; Nazarov et al., 2012; Agatova et al., 2014b). Table 2 presents all available at the moment radiocarbon dates of wooden fragments at the heads of glaciated valleys within the mountain ranges of the SE Altai. Figures 4 and 5 summarize the studied radiocarbon chronology of climatically driven mountain glaciers advances and retreats in the SE Altai, which is discussed below after presenting the main ideas of interpretation and utilization of obtained absolute ages.

Glaciological interpretations of dated samples

Dating of glacially-generated landforms requires corresponding attention to the processes leading to the formation or deposition of the dated features. The interpretation of data derived from collected samples depends on their history. Sources of uncertainty include different initial locations of wood, transportation, and the position of the wood at collection sites. Therefore, it is important to specify the history of the samples in order to enable a conclusive interpretation.

Our reconstructions are based on hypothesis that the date of a tree's death by itself argues for favorable conditions for living before this date rather than for adverse conditions leading to their death. The only exception is the youngest Aktru stage where the researcher had the opportunity to confirm that the death of the tree was caused directly by an advancing glacier.

One of the arguments supporting this hypothesis is that besides glacier activity at the head of the valleys, the tree's death can be caused by many reasons: rock falls, landslides, debris flows, snow avalanches, forest fires etc. Intensification of slope processes in a periglacial area of trough valleys is controlled by various factors from climatic (rain and snowfalls, snowmelt, floods) to seismic ones. The recurrence interval of strong earthquakes in the framing ranges of the Kurai-Chuya system of intermountain depressions in the second half of the Holocene is about 400 years (Agatova et al., 2014a), so in this area strong earthquakes are more frequent than intensive cooling. The 2003 Chuya earthquake (M_S=7.3) caused not only ruptures on the glaciers, but also rock falls and death of trees in periglacial areas. Strong paleoearthquakes could have caused the seismogravitational deformations of moraine ramparts and tree's death on their periphery, which could be falsely interpreted as a time of maximal glacier advances. Finally, mass tree deaths and burial at the foot of steep slopes caused by intensification of slope processes can occur during glacier recession (Ivanovsky et al., 1982).

Table 2. Available radiocarbon dates of wooden fragments and charcoals which characterize the Holocene glaciers fluctuations in mountain ranges of the SE Altai

Lab code	Sample type [1)]	Sample location (Figure.2) [2)]	Altitude a.s.l. (m)	^{14}C Age [3)]	Calibrated (years BP) [4)]	Number of rings	Reference
SOAN 1616	1	A		430 ± 50	482 ± 57	-	Ivanovsky and Panychev, 1978
SOAN 1617	2	A		610 ± 50	601 ± 64	-	
SOAN 1618	1	A		440 ± 50	485 ± 59	-	
SOAN 1426	1	A		430 ± 30	490 ± 39	-	
SOAN 1427	1	A		430 ± 30	490 ± 39	-	
SOAN 1881	4	K		6100 ± 50	7006 ± 154	-	Ivanovsky, 1993
SOAN 1882	4	K		6080 ± 60	6913 ± 120	-	
SOAN 1883	4	K		2920 ± 35	3063 ± 104	-	
SOAN 1884	4	K		3075 ± 25	3301 ± 62	-	
SOAN 1885	1	K		465 ± 30	514 ± 26	-	Ivanovsky et al., 1982
SOAN 1887	1	A		130 ± 20	105 ± 46	-	
SOAN 1888	1	A		125 ± 15	103 ± 44	-	
TA - 859	4	K		6060 ± 60	6903 ± 128	-	Okishev, 1982
SOAN 4116	4	S		2630 ± 60	2780 ± 90	-	Agatova et al., 2002
SOAN 4117	1	S		710 ± 35	676 ± 37	-	
LU 1526	1	D		520 ± 80	549 ± 118	-	Mikhailov, 1987
LU 1829	1	K		3710 ± 40	4057 ± 99	-	Mikhailov et al., 1989
LU 1831	1	K		3800 ± 70	4238 ± 175	-	
Hv 24148	4	P	1755	910 ± 145	890 ± 251	-	Schlütz and Lehmkuhl, 2007
Hv 24149	4	P	1755	1070 ± 155	1005 ± 281	-	
Hv 24150	4	P	1755	1010 ± 155	932 ± 255	-	
Hv 24151	4	P	1755	1170 ± 45	1075 ± 106	-	
SOAN 5449	2	A	2188	580 ± 55	575 ± 54	60	Agatova et al, 2012

Table 2. (Continued)

Lab code	Sample type [1]	Sample location (Figure.2) [2]	Altitude a.s.l. (m)	^{14}C Age [3]	Calibrated (years BP) [4]	Number of rings	Reference
SOAN 5450	2	A	2173	615 ± 50	602 ± 63	60	Agatova et al, 2012
SOAN 5631	1	A	2242	2855 ± 80	2999 ± 213	140	
SOAN 5632	1	A	2242	2770 ± 80	2912 ± 167	40	
SOAN 5633	1	A	2324	785 ± 80	738 ± 98	-	
SOAN 5634	1	A	2344	670 ± 85	630 ± 111	30	
SOAN 5635	1	A	2331	1105 ± 85	1045 ± 143	50	
SOAN 6033	3	M	2234	1525 ± 45	1427 ± 98	200	
SOAN 6034	3	M	2202	850 ± 45	743 ± 61	50	
SOAN 6035	1	M	2218	915 ± 45	831 ± 92	-	
SOAN 6036	1	M	2218	775 ± 45	719 ± 58	-	
SOAN 6037	3	M	2399	1120 ± 70	1052 ± 132	370	
SOAN 6038	3	K	2327	665 ± 35	651 ± 26	-	
SOAN 6039	3	K	2310	645 ± 35	583 ± 30	250	
SOAN 6040	1	A	2332	730 ± 45	688 ± 49	-	
SOAN 6041	1	A	2234	4315 ± 85	4930 ± 145	-	
SOAN 6042	1	A	2249	1205 ± 25	1121 ± 61	-	
SOAN 6454	3	H	2462	6175 ± 125	7095 ± 231	40	
SOAN 6455	2	H	2281	590 ± 65	592 ± 73	-	
SOAN 6456	1	H	2173	1140 ± 70	1057 ± 128	50	
SOAN 6457	4	K	2384	4350 ± 45	4914 ± 72	50	
SOAN 6458	2	H	2155	470 ± 60	495 ± 68	120	
SOAN 6459	2	H	2160	480 ± 90	498 ± 77	400	
SOAN 6460	3	H	2165	400 ± 70	420 ± 113	100	
SOAN 6461	3	H	2160	435 ± 60	484 ± 64	50	
SOAN 6462	2	H	2138	485 ± 60	498 ± 69	140	

Lab code	Sample type [1]	Sample location (Figure.2) [2]	Altitude a.s.l. (m)	^{14}C Age [3]	Calibrated (years BP) [4]	Number of rings	
SOAN 6463	1	H	2208	3650 ± 75	3990 ± 170	50	
SOAN 6464	3	H	2310	450 ± 60	490 ± 68	100	
SOAN 6465	3	H	2397	1185 ± 70	1116 ± 151	100	
SOAN 6466	2	H	2198	385 ± 35	466 ± 43	150	
SOAN 6467	1	M	2176	5230 ± 55	6018 ± 112	-	
SOAN 6468	1	M	2158	4890 ± 50	5658 ± 78	-	
SOAN 6469	3	A	2377	1515 ± 75	1421 ± 127	-	
SOAN 6470	1	A	2240	4840 ± 80	5593 ± 147	-	
SOAN 6471	2	A	2207	465 ± 70	493 ± 72	140	
SOAN 6472	2	A	2222	520 ± 90	549 ± 121	-	
SOAN 6473	4	K	2384	4500 ± 50	5171 ± 140	-	
SOAN 6911	2	H	2157	460 ± 40	501 ± 48	150	
SOAN 6912	2	H	2155	480 ± 35	518 ± 33	50	
SOAN 6913	2	H	2199	435 ± 35	486 ± 49	-	
SOAN 6914	2	H	2200	535 ± 40	536 ± 29	230	
SOAN 7376	1	M	2156	4985 ± 95	5753 ± 170	70	
SOAN 7377	1	M	2158	4680 ± 90	5433 ± 166	450	
SOAN 7378	1	M	2173	4710 ± 100	5461 ± 190	-	
SOAN 7379	1	A	2218	710 ± 55	679 ± 56	-	
SOAN 7380	1	A	2261	700 ± 50	677 ± 52	100	
SOAN 7381	1	A	2260	1000 ± 50	889 ± 101	-	
SOAN 7382	1	A	2260	690 ± 45	659 ± 39	-	
IGAN 3645	1	A	2234	4640 ± 100	5317 ± 271	-	
IGAN 3702	1	A	2236	4810 ± 60	5551 ± 103	-	
IGAN 3875	1	M	2157	4870 ± 70	5607 ± 143	280	
IGAN 3878	3	M	2196	500 ± 60	513 ± 59	140	
IGAN 3879	1	H	2149	3450 ± 70	3726 ± 170	100	

Table 2. (Continued)

Lab code	Sample type [1]	Sample location (Figure.2) [2]	Altitude a.s.l. (m)	^{14}C Age [3]	Calibrated (years BP) [4]	Number of rings	Reference
IGAN 3880	1	M	2174	4540 ± 80	5146 ± 187	-	
IGAN 3886	3	M	2182	380 ± 60	411 ± 102	460	
SOAN 6915	1	S	2451	3525 ± 70	3846 ± 214	40	Nazarov and Myglan, 2012
SOAN 6916	1	S	2459	3365 ± 95	3618 ± 221	50	
SOAN 6925	1	R	2149	5595 ± 85	6419 ± 149	120	
SOAN 6926	1	R	2137	5545 ± 85	6358 ± 176	-	
SOAN 6927	1	R	2137	5315 ± 75	6077 ± 143	90	
SOAN 6930	1	R	1970	5395 ± 75	6151 ± 161	-	
SOAN 6931	1	R	1995	2490 ± 75	2549 ± 187	270	
SOAN 7383	1	S	2480	4270 ± 110	4797 ± 276	-	
SOAN 7386	1	R	1986	4610 ± 80	5264 ± 221	60	
IGAN 3694	1	S	2466	4050 ± 80	4563 ± 264	-	
IGAN 3695	1	S	2466	3550 ± 110	3858 ± 292	100	
IGAN 3696	1	T	2483	5440 ± 80	6200 ± 198	100	
IGAN 3697	1	Q	2100	4660 ± 70	5330 ± 256	-	
IGAN 3704	2	Q	2221	960 ± 70	889 ± 161	250	
IGAN 3876	1	R	2016	2540 ± 70	2562 ± 198	100	
IGAN 3877	1	R	2147	4520 ± 60	5205 ± 233	-	
IGAN 3882	1	S	2458	3040 ± 70	3198 ± 193	100	
IGAN 3883	1	R	2016	1440 ± 70	1356 ± 164	-	
IGAN 3884	1	R	2016	2310 ± 70	2410 ± 281	100	
IGAN 3887	1	S	2476	3020 ± 70	3188 ± 185	-	

Lab code	Sample type [1)]	Sample location (Figure.2) [2)]	Altitude a.s.l. (m)	^{14}C Age [3)]	Calibrated (years BP) [4)]	Number of rings	References
SOAN 6917	2	Q	2106	400 ± 40	471 ± 47	300	Galahov et al., 2012
SOAN 6918	2	Q	2114	1400 ± 70	1296 ± 122	200	
SOAN 6919	1	Q	2113	3600 ± 85	3896 ± 207	70	
SOAN 6920	1	Q	2109	2895 ± 35	3043 ± 118	40	
SOAN 6921	1	Q	2124	3140 ± 30	3385 ± 59	120	
SOAN 6922	1	Q	2095	2970 ± 45	3132 ± 135	150	
SOAN 6923	1	Q	2100	3835 ± 45	4278 ± 133	-	
SOAN 7384	3	Q	2225	530 ± 50	535 ± 33	250	
SOAN 7385	3	Q	2225	270 ± 70	379 ± 123	200	
SOAN 7388	1	R	1995	4800 ± 80	5548 ± 114	20	
IGAN 4089	4	B	2470	7640 ± 100	8415 ± 212	-	Agatova et al, 2014a
IGAN 4092	4	T	2140	1460 ± 70	1400 ± 120	-	Agatova et al, 2014b
IGAN 4139	4	I	1755	1530 ± 70	1426 ± 125	-	
IGAN 4140	4	I	1755	1290 ± 70	1192 ± 135	-	
SOAN 8417	4	T	2140	2755 ± 40	2859 ± 88	-	
SOAN 8658	4	T	2140	650 ± 65	611 ± 78	-	

[1)] Sample type: 1- buried wood fragments in moraines, glaciofluvial sediments and proglacial forefields; 2- trees died "*in situ*" in front of moraines and partly covered by moraines; 3- wood fragments and tree trunks on the moraines surfaces and on trough slopes; 4- wood fragments and charcoals buried by slope sediments.

[2)] Sample locations are shown in Fig. 2:

A- Aktru valley; B- Boguty valley; D- Berel valley; H- Shavla valley; I- Chuya Valley, mouth of the Kuektanar river; K- Kurkurek, Jan-Karasu valleys and Ehtykkel stow; M- Maashey valley; P- foot of the Kurai range; Q- Valleys of Akkem and Kucherla rivers; R- Mensu valley; S- Akkol valley and Sofiyskiy glacier; T- Taldura valley.

[3)] ^{14}C age is conventional radiocarbon age.

[4)] Calibrated ages are given at the 2-σ level.

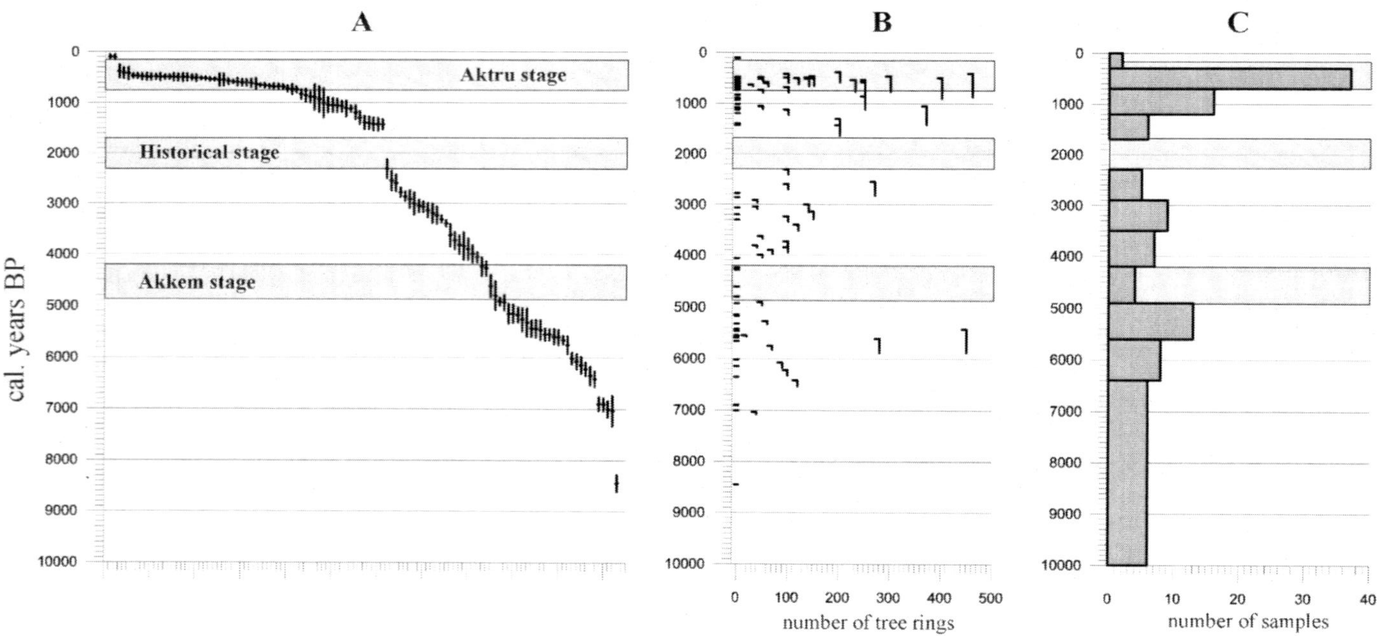

Figure 4. The Holocene glacier dynamics forced by climate changes in the SE Altai. A) Periods of glacier advances (which extended the size of modern glaciers) based on 113 radiocarbon ages (Table 2). B) Number of tree rings preserved for each dated wooden fragment. Each sample represents a certain time interval of tree growth in the heads of the valleys. This interval is defined by the lifetime of the plant before its death. C) Histogram counting the number of samples per period (the durations of the periods is listed in Table 3) using the median calibrated age.

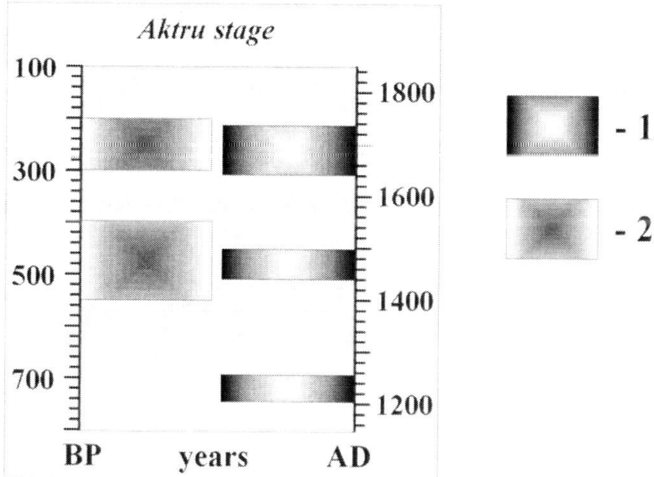

Figure 5. Periods of climate deterioration within Aktru stage (grey shaded) specified by 1) dendrochronological analysis and by 2) ice core drilling on Belukha peak, Katun range, after (Agatova et al., 2014b).

Another argument is that the time period between a tree's death and its burial by glacier is usually undefined. For example in SE Altai, *Larix Sibirica* can be preserved on a stone surface up to two thousand years (Agatova et al., 2014a). Moreover, single ages of dead trees toppled by moraines cannot be directly correlated with the beginnings of a glacial advance. Only a large amount of trees buried in different valleys during a short time period indicates the beginning of a glacial stage. Progressive glacier advance and successive death of trees could be evidenced only by findings of tree fragments "in situ" (Le Roy et al., 2015), however, such findings are generally rare.

The extended dataset of favorable periods for tree living in heads of troughs - above the modern upper tree limit on valleys slopes and in the modern glaciated area, limits the times of climate deterioration and possible glacier advances in a complementary way, and thus glacial stages are marked between these dated periods.

It should be also mentioned that response characteristics of mountain glaciers vary significantly from tens to more than one hundred years (Haeberli and Hoelzle, 1995). The length response time (the parameter which characterizes the glacier's steady-state sensitivity to sudden massbalance changes) and the reaction time (which is defined as the time lag between a sinusoidal climate forcing and the pseudosinusoidal front variation response of a glacier) for Sofiysky and Maly Aktru glaciers (Russian Altai) are about one hundred years (De Smedt and Pattyn, 2003; Agatova and Nepop, 2010). Thus decadal climate fluctuations make no significant influence on the SE Altai glacier behaviors and their evolutions (advance/retreat) will instead follow the dominant climatic trend (centennial-scale and longer).

Archeological data from Scythian burials associated with the Pazyryk culture in Chuya intermountain depression give additional information about prevailing climate and forest expansion in southeastern part of Russian Altai, and thus could be indirectly utilized for determining the periods of climate deteriorations which controlled glaciers expansion in the region. *Larix Sibirica*, utilized for wooden burial constructions in the source of the Chuya river and the valleys of its tributaries - Ulandryk, Justyd, Barburgazy, Sebistei (eastern part of Chuya intermountain depression), has been found to have grown during 720-297 years BC from dendrochronological analysis (Slyusarenko, 2010). Calendar dating of the floating 424-years tree-ring chronology was carried out by high resolution radiocarbon dating, applying of the wiggle-matching procedure (Slyusarenko et al., 2001; Slyusarenko, 2010), and by cross-dating with the 2367-years absolute tree-ring chronology constructed for the SE Altai and Tuva regions (Myglan et al., 2012). The wide distribution of *Larix Sibirica* , which is evidenced by considerable timber usage of the Pazyryk population at that time (720-297 years BC), implies more humid and warm climate conditions favorable for forest growth.

Chronology of Climatically Driven Glaciers Fluctuations

Newly obtained extensive dataset refutes the traditional concept of the Russian Altai Holocene glaciations as a consecutive retreat of the late Würm glaciers and argue their practically complete degradation at the beginning of the Holocene. In the second half of the Holcene, during the Neoglacial, glaciers repeatedly advanced and at least three times they expanded beyond the size of modern glaciers. During interglacial worming glaciers practically totally degraded or their size didn't exceed the size of modern ones.

Radiocarbon ages of plant remains in lacustrine sediments from a small Dzhangyskol lake of glacial origin and from pingos in Eshtykkel stow at about 1750 m a.s.l. (Blyakharchuk et al., 2008) (Figure 2 K) confirm deglaciation of the foot of the north slope of the North Chuya range had occurred by 15000 cal. years BP. Our geomorphological investigations and radiocarbon dating of fossil soils covered moraines in the Boguty river valley (the eastern flank of the Chuya basin, Figure 2 B) at 2470-2540 m a.s.l. indicate that by 14000 cal.

years BP Late Würm (Sartan) glaciers had retreated at least above 2600 m a.s.l. in trough valleys of the Chikhachev range (Agatova et al., 2014a).

By the early Holocene a more humid and warm climate in comparison with the modern one dominated in the region. Significant glaciers degradation within the SE Altai and neighboring SW Tuva is evidenced by finds of wooden fragments (10000-11000 cal. years BP) above the modern timber limit (Nazarov et al., 2012). Late Würm (Sartan) glacier retreating in mountain valleys of the Chikhachev and South Chuya ranges and a prolonged period of soil formation is also demonstrated by the radiocarbon ages of buried soil at altitudes of 2000-2320 m a.s.l. in Boguty (Butvilovsky, 1993; Agatova et al., 2012) (Figure 2 B), Tara (Butvilovsky, 1993) (Figure 2 L) and Chikty (Schlütz and Lehmkuhl, 2007) (Figure 2 C) river valleys. Modern soils at this altitude are less humus or are totally absent. In contrast to the modern poor grass vegetation, the abundance of charcoal in fossil soil argues for the expansion of forest vegetation, which is not now presented in these valleys. Thus soil formation and forest growth indicate a significant glacial retreat by 10000 cal. years BP, and further glacier disintegration and ice melting in crest areas of the ridges.

The first half of the Holocene in the SE Altai is generally characterized by significantly more humid and warm climate conditions in comparison to the modern ones. This is evidenced by soil development in mountain valleys (Butvilovsky, 1993; Agatova et al., 2014a) as well as by finds of *Larix Sibirica* fragments in the sediments of rock glaciers in the Jan-Karasu valley, North Chuya range (Figure 2 K) (Ivanovsky, 1993). The last fact testify that about 7000 BP the upper timber limit on the northern slope of the North Chuya range was about 300 m above its modern position.

The broad range of radiocarbon dates for numerous wood fragments melted out from glacier tongues and found on trough slopes above the modern upper timber limit in various glacial valleys within the SE Altai argue for steady warming about 6000-4900 cal. years BP (Agatova et al., 2012). At this time, forest vegetation occupied even the heads of troughs. The death of the trees found could be caused by either climatic or age reasons.

Possible periodic cooling accompanied by glacier expansion could occur in the first half of the Holocene but by now total amount of absolute dates of paleowood related to this time period doesn't allow reliable reconstructing climate and glacial events.

The absence of tree finds with the age 4900-4200 cal. years BP could be attributed to abrupt climate deterioration about 5000 cal. years BP. Favorable conditions for tree existence during this period could have been reduced due to cooling. Most likely formation of "Akkem" moraines in the main glacial valleys of the North Chuya range could be correlated with this climate deterioration. At the present time, there are no absolute dates directly connected with the ages of these moraines. Nevertheless the ages of peat and lacustrine sediments covering the "Akkem" moraine in Aktru valley (North Chuya range, Figure 2 A) indicate its formation before 3000 cal. years BP (Agatova et al., 2012). Absence of plant remains of wood or bush vegetation at the foot of the lacustrine sediments covering the "Akkem" moraine here suggests a relatively short time interval between moraine surface stabilization and lake formation. A thin layer of sediments that had been accumulated on the bottom of the lake also argues for the Holocene age of the "Akkem" moraines and proposes their formation during the period 4900-4200 cal. years BP, as characterized by absence of tree finds in the heads of glacial valleys (Agatova et al., 2012).

As it was mentioned above paleoglaciological reconstructions related to the Akkem stage are the most debatable. The absence of direct absolute age estimations of associated moraine complexes as well as location of these moraines far downstream from modern glaciers and their presence not in all mountain valleys form the basis for discussions. Correlation of the most distant glacial landforms in mountain valleys of some ranges of the SE Altai with climate deterioration during the Akkem stage (Nazarov and Agatova, 2008; Agatova et al., 2012; Nazarov et al., 2012; Agatova et al., 2014a,b) implies that formation of huge glaciers many kilometers in length was very rapid - during several hundred years. The abrupt glacier advances during this stage were most probably caused by the coincidence of extremes of lowered temperature and increased precipitation. Degradation of "Akkem" glaciers was also very rapid. The largest glaciers left extensive fields of dead and buried ice, thermokast lakes, and swales in the firn valleys whilst retreating. In this case strongest and most intensive Holocene glacier advance amongst all those expressed in the topography lasted less than thousand years.

Further warming at 3300-2300 cal. years BP and glaciers retreating in the mountain ranges of the SE Altai is evidenced by soil formation in the eastern part of the Chuya intermountain depression, in the Chuya valley between Chuya and Kurai depressions, within the North and South Chuya ranges and by spreading the forest vegetation beyond the modern upper timber limit as well as by considerable width of tree rings (1-1.5 mm - at least two times larger than the width of modern ones, which sometimes are formed by only several cells) of *Larix sibirica* which grew in the modern glacial zone in the North Chuya range. Intensification of slope processes in the mountain valleys of the North and South Chuya ranges and alluvial fans development within the Kurai intermountain depression argue for an increase in precipitation at around 3000 cal. BP which accompanied this warming (Agatova et al., 2012).

Dendrochronological analysis of numerous finds of *Larix sibirica*, utilized for wooden burial constructions in the source of Chuya river and the valleys of its tributaries in the eastern part of Chuya depression (Figure 2 J, U), demonstrates the growth of trees at 2670-2247 BP (Slyusarenko, 2010). Using larch trunks for burial of ordinary members of society in 14 kurgans of the family burial ground "Ulandryk1" (Figure 2 U) suggests a broad expansion of *Larix Sibirica* at now arid and desert areas of the mountain framing of the Chuya intermountain depression. An archaeological study of 42 Scythian burials within the Ulandryk valley (Figure 2 U) argues for considerable timber usage for burials chambers which on average were about 2.2 x 1.5 m in size and 0.5-1.0 m in height. They were constructed from massive larch logs up to 0.5 m in diameter and even more and had "roofs" of 4-6 wooden blocks and 5-6 well hewn floor boards (Kubarev, 1987). Local inhabitants also reported finds of single buried stubs of larch in the head of Aksai river (Figure 2 N) and the name of Justyd River located 5-10 km northeast means "hundred larches" in Altai language. Thus, wide distribution of *Larix Sibirica* at that time (2670-2247 BP) implies more humid and warm climate conditions favorable for forest growth, even in the arid and desert eastern part of Chuya intermountain depression.

Tree ring chronology of Pazyryk burials in the Chuya depression (Slyusarenko, 2010) reveals that cooling started about 2300 cal. BP. This fact is also supported by cryoturbation of loamy sand and soil layers from an alluvial fan of the Kuraika river (Figure 2 E). Further climate deterioration led to the formation of an area of permafrost soil within the Kurai depression and is responsible for glacier expansion within framing ridges during the subsequent Historical stage.

At the present time there are no direct radiocarbon dates for "Historical" moraines in the SE Altai. On the basis of dating warm periods before and after the Historical stage, its chronological limits are estimated from about 2300 until 1700 cal. BP. Steady reduction in tree ring growth from 2300 until 2000 cal. BP is also recorded in the "Mongun" tree ring chronology, which shows the climatically driven radial tree growth in the SE Altai and neighboring Tuva region (Myglan et al., 2012). Complex structure of "Historical" moraines (up to three oscillation ramparts were formed at the heads of the valleys) indicates alternating periods of warming and cooling during this stage.

The beginning of steady warming about 1700-1600 cal. years BP which is evidenced by recolonization of the "Historical" moraine and tree settlement above the modern upper tree limit in some valleys of the North Chuya range can be regarded as the end of Historical stage (Agatova et al., 2012). Further large-scale forest expansion beyond the upper tree limit indicated significant glacier retreat within the SE Altai. During this warm period before the Aktru stage (LIA) trees grew in a modern glacial zone. The heads of the valleys were occupied by forest vegetation which was later killed by cooling and glaciers advances during the Aktru stage. By 7th century soil had been formed in the mouth of the Kuektanar river (Figure 2 I). The second half of this warm period (9th – 12th centuries) is characterized by a maximum number of tree fragments located beyond the modern upper tree limit. Short-term abrupt decrease of tree ring growth is recorded in "Mongun" tree ring chronology in 6th century, but until the beginning of 13th century possible minor glacier oscillations did not exceed the advances of the Aktru stage.

Numerous radiocarbon dates of trees toppled by moraines argue that the earliest glacier advances during the Aktru stage took place in the second half of the 13th century. The results suggest that the larger glaciers started to advance first. Activation of the smaller glaciers began at the second half of 15th – 16th centuries (Agatova et al., 2012). During the Aktru stage there were several glacier oscillations which are morphologically expressed and confirmed by lichen analysis (Solomina, 1988). Dendrochronological analyses reveal periodic forest renewals in the first half of the 15th, in the middle of the 16th, and at the beginning of the 17th centuries at the head of valleys (Dushkin, 1965; Adamenko and Sjubaev, 1977; Adamenko and Selischev, 1984). This fact also supports the hypothesis that the LIA was not a single cold period, but instead consists of several alternating cold and warmer time intervals (Solomina, 1999). New dendrochronological dates (Agatova et al., 2014 b) of death for trees grown even below the upper timber line fall on time periods AD 1206-1256, 1445-1501, 1642-1736 (Figure 5). Last two of them coincide with the periods of cooling revealed by ice drilling on Belukha peak (Katun range) - AD 1400-1550 and 1650-1750 (Nazarov et al., 2012). Based on the radiocarbon dating and dendrochronological analysis, the time limits of the Aktru stage could be defined as from the second half of the 13th century until the end of the18th - first half of the 19th centuries.

Since the second half of the19th centuries till now the SE Altai mountain glaciers have been permanently degraded. The thermal minimum in the middle of the 19th century, the greatest in the last millennium, did not positively influence the mass balance of the glaciers. Since AD 1898, when Sapozhnikov discovered and described many Altai glaciers (Sapozhnikov, 1912), they have subsequently retreated with an average rate of 10-18 m a^{-1} (Agatova et al., 2002; Pattyn et al., 2003), remaining in a near steady state during the beginning (about AD 1910) and in the middle of the 20th century (Bulatov et al., 1967).

Presented data argue for a hypothesis of periodic but successively less Holocene glacier expansion from about 5000 cal. years BP to the Little Ice Age, presumably driven by aridity intensification in the Altai throughout the Holocene. A decrease in the areas occupied by forest vegetation and the reduction in glacier size at each succeeding glacial stage suggest an increasing moisture deficit in this mountain province during the second half of the Holocene. This conclusion is also supported by the absence of a reaction from the glaciers to the thermal minimum of the middle of 19th century. It can be seen that the role of aridity (as well as temperature) as climatic forcing factors in driving glaciation in the Altai Mountains, as well as in the whole Central Asia, cannot be overestimated.

Quantitative Assessment of Glacial Erosion Rate, Response Characteristics of Modern Glaciers and Development of Longitudinal Profile of Glacial Valleys in the SE Altai on the Basis of Numerical Simulation

In previous section we have demonstrated the presence of three Holocene glacial stages within the SE Altai when mountain-valley glaciers were larger than modern ones. Significantly more powerful Pleistocene climate deteriorations determined the development of much more extensive semi- ice-sheet and valley-piedmont glaciations in the region (Devyatkin, 1965; Okishev, 1982; 2011; Agatova, 2005; Zol'nikov et al., 2010). These time periods characterize much more intensive relief-changing glacial activity which can be generally evaluated by estimating glacial erosion rate. In this section we present one-dimensional numerical model of idealized glacier which describes the ice flow along a central flow line. Applying numerical simulation of the ice movement and its interaction with the glacier bed allows quantifying the glacial effects on topography changes within the SE Altai..

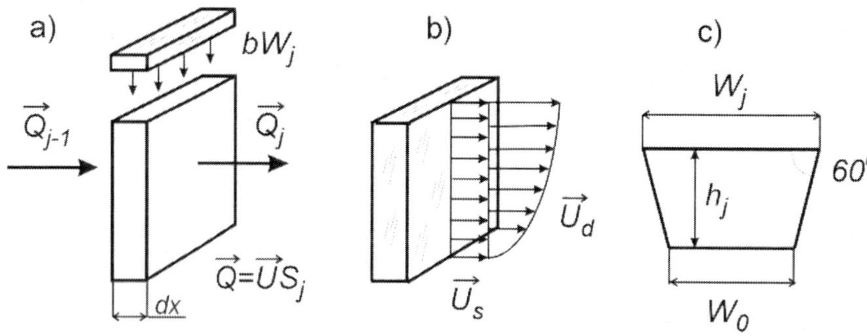

Figure 6. Schematic diagram of numerical model cell. *a* – ice discharge across cell edges; *b* – vertical profile of ice velocity at central flow line; *c* – valley cross section profile.

Statement of a Problem for Numerical Simulation

Glacier activity is a quite complicated process. Moving ice is a polycrystalline substance consisting crystals of different shapes, sizes and orientations. It contains impurities, meltwater and transports rock fragments of various shapes and sizes. Ice temperature varies in a wide range. There are also different basal conditions. Glacier erosion activity includes many processes: glacial abrasion, plucking, exaration, glaciofluvial erosion which proceed simultaneously. Intensity of glacier erosion activity is controlled by glaciological (temperature of the basal ice layers, ice rheology etc.), geological geomorphological (lithology, fracturing of bed rocks, bed topography etc.), and climatic factors.

As a result of this erosion impact, mountain valleys with specific longitudinal and transverse profiles – troughs, are developed. At the same time the climatic zonation and tectonic movements, which play especially important role in a long time interval, as well as peculiarities of erosion transformation of the valleys during interglacial periods cause a plenty variety of glacial valleys even within the same mountain system. Solving the inverse problem – reconstruction of the topography evolution on the basis of modern relief morphology, is often associated not only with the problems of establishing functional relations between different geological processes during the time, but also with recognizing the processes themselves involved in production and transformation of different landforms. Thus in our researches we have solved direct one-dimensional problem which describes the motion of idealized glacier and its interaction with the valley bottom. Some simplifications were adopted in numerical simulation.

First of all, the numerical model takes into account only exarational and abrasive glacier activity, and doesn't considerate fluvioglacial processes on the glacier bed. As a result, calculated value is the lowest assessment of the glacial erosion rate, and obtained longitudinal profiles are more shallow and smooth than those produced under the real nature conditions.

Secondly, the accumulative glacier activity is not considered. Debris material transported by a real glacier is deposited in the low part of the valley. It determines the development of the longitudinal valley profile and therefore ice movement near the glacier terminus. Applying this simplification the suggested model gives the most accurate estimates of erosion rate for upper and middle sections of glacial valley.

Thirdly, the model does not take into account the possible presence of water in the basal ice layers, which affects the shear stress on the glacier bed. Results of studying the basal conditions for Sofiysky glacier (South Chuya range, Figure 2 S) are presented in (Pattyn et al., 2003).

Finally, numerical simulation was carried out for constant climatic conditions, relevant to the modern ones.

Numerical model of glacier ice flow

Glacier movement is described in terms of changing the ice thickness - h along the central flow line during in time (Figure 6):

$$\frac{\partial Q}{\partial x} + \frac{\partial S}{\partial t} = bW,$$

where $Q = US$, $U = \overline{U}_d + U_S = \frac{2}{5} Ah\tau_b{}^3 + C\frac{\tau_b{}^3}{\rho gh}$, $\tau_b = F\rho gh \sin\alpha$.

Here Q is ice flow through the cross section of the glacier S; U – averaged over the depth ice velocity along the central flow line; $\overline{U}_d$ – average deformation rate; U_S – slipping rate; τ_b – share stress on the bedrock; b – glacier mass balance, averaged over the width of the glacial valley W; g – gravitational acceleration; α – bedrock inclination (Greuell, 1992; Paterson, 1994); x–axis is directed along the bedrock downstream the valley.

A two-dimensional geometry of a glacial valley transverse profile is taken into account by introducing an additional coefficient F - shape factor. Its value is assumed to be 0.8, which is calculated according to the algorithm suggested in (Paterson, 1994) for a trapezoid transverse valley profile with a base angle of 60°. Initially the longitudinal profile of a glacial valley is assumed to be linear. Real glacial valleys have the bottom areas with over-deepening or step-type bedrock profiles, where erosion is more intensive (Hooke, 1991). Thus calculated in this work, longitudinal profiles are more smooth in comparison with the real ones.

Sliding coefficient - C and Arrhenius constant - A are taken $0.8 \cdot 10^{-16}$ Pa^{-3} a^{-1} and $1.0 \cdot 10^{-8}$ m Pa^{-1} a^{-1} accordingly (De Smedt and Pattyn, 2003); ice density - 870 kg m^{-3} (Paterson, 1994). Linear correlation between surface mass balance and altitude, calculated on the basis of 1962-1999 instrumental observations on the Maly Aktru glacier and 1998-2000 surface mass balance measurements on the Sofiysky glacier is adopted (Pattyn et al., 2003) (Figure 7).

It was implied that erosion rate - ε in every point of the valley bottom is proportional to the glacier slipping rate (Hallet, 1979; 1996):

$$\varepsilon = C_\varepsilon \, U_S,$$

where $C_\varepsilon = 1.0 \cdot 10^{-4}$ (Humphrey and Raymond, 1994).

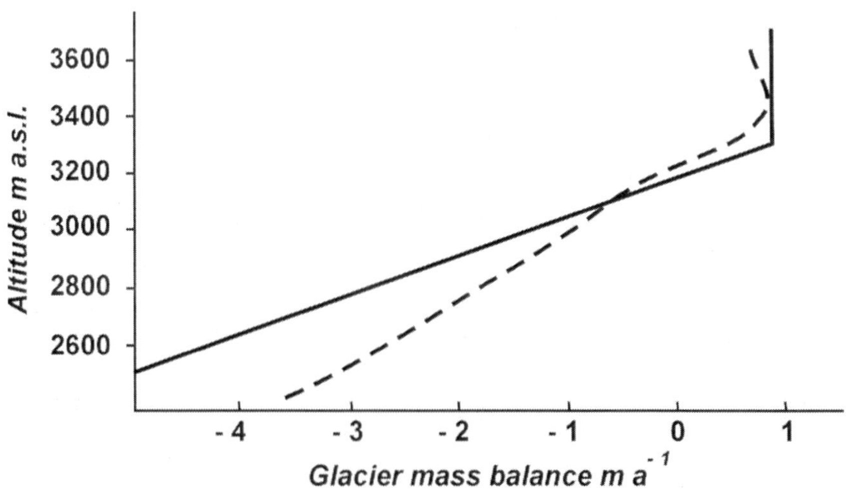

Figure 7. Glacier mass balance reference profiles after (Pattyn et al., 2003) for 1) Sofiysky and 2) Maly Aktru glaciers.

Glacier with an ice thickness about 100 m (maximal ice thickness 120 m) and total length 5 km, located on the inclined plane with maximum and minimum altitudes 3500 m and 2600 m respectively, was considered as a model one. These characteristics are similar to parameters of the longitudinal profile along the central flow line of the Sofiysky glacier (Pattyn et al., 2003). As it is shown below under stable climate conditions at the time interval of about 10^2 years the model glacier goes into a stationary state with the typical for real glaciers surface shape. Thus initial "arbitrary" surface shape of a model glacier on the time scale of about 10^3 - 10^5 years doesn't affect the calculations of erosion rate and evolution of the longitudinal valley profile.

Numerical Simulation Results and Their Comparison with the Results of Other Studies

Numerical simulation included three series of numerical experiments: calculating glacier erosion rate, qualitative estimation of bedrock profile evolution, and calculating characteristic time of glacier response to climate change (Agatova and Nepop, 2010) (Figure 8).

Absolute glacier erosion rate under the modern climate conditions of the SE Altai on the basis of numerical simulation is about $1.0 \cdot 10^{-3}$ m a^{-1} with its maximal value about 2.0 m a^{-1} near ELA position. This parameter calculated for the Maly Aktru glacier by Dushkin (1974) is one order of magnitude lower - $(0.13\text{-}0.15)10^{-3}$ m a^{-1}. This assessment was based on field measurements of ice velocity and quantitative content of moraine material in the glacier terminus. The disadvantage of this technique is overestimating the erosion rate due to the debris volume coming from the valley slopes. In this case Maly Aktru glacier was considered as a reference glacier because it differs from other mountain glaciers of the North Chuya range by insignificant volume of surface moraines. These estimates - $(0.13\text{–}0.15)10^{-3}$ – $(1.0\text{–}2.0)10^{-3}$ m a^{-1}, most likely limit the range of possible values of erosion rates for modern valley glaciers of the SE Altai. They are defined by regional glacier characteristics which, in its turn, are controlled by local oroclimatic features.

The magnitude of glacial erosion rate, calculated on the basis of estimating the energy of moving ice, indicates that several millimeters per year of bedrock destruction by moving glacier is quite possible. Moreover intensity of exaration can not be measurement by values less than 10^{-1} millimeters per year for unbroken solid rocks (Chernova, 1974).

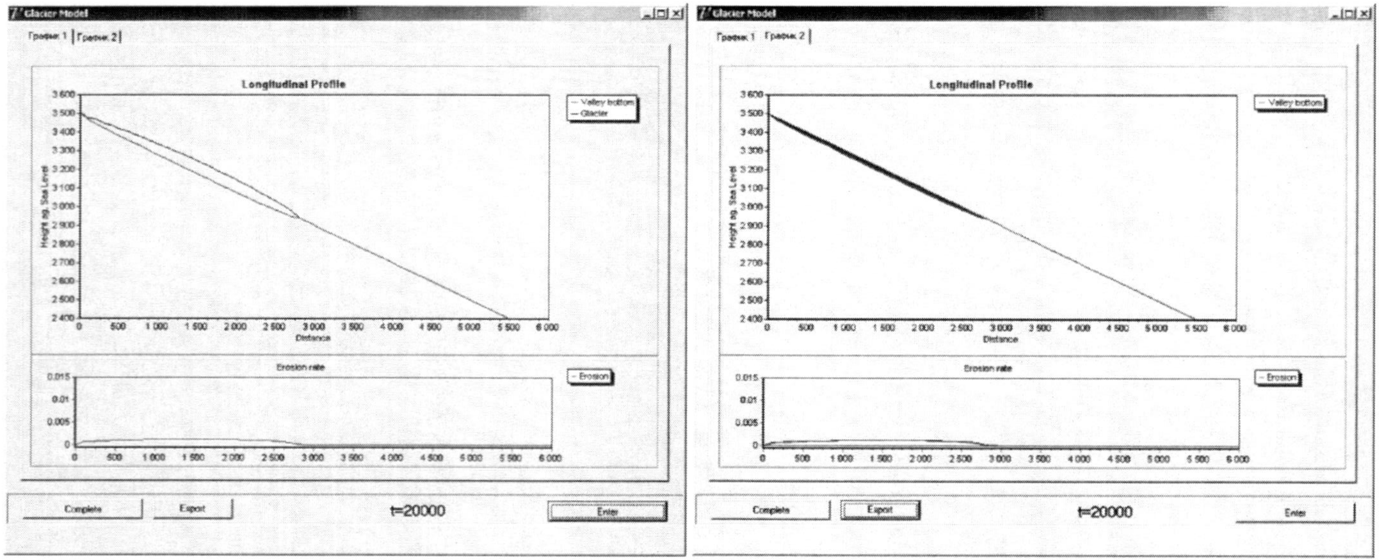

Figure 8. Computer program that simulates ice movement along the central flow line in the glacier and its interaction with the valley bottom. Glacier position and evolution of the longitudinal profile are shown in the upper chart of the left and right windows correspondingly; glacier erosion rate – in the lower chart. Spatial data are given in meters, time in years and erosion rate in m a^{-1}.

These estimates are also supported by reviewed field data on rates of glacial erosion and sediment evacuation by glaciers presented in (Hallet et al., 1996). Erosion rate of modern mountain-valley glaciers is reported to be limited between $0.01 \cdot 10^{-3}$ and $10 \cdot 10^{-3}$ m a^{-1}.

Table 3. Stages of glaciers advances (grey shaded) and interglacial periods in the SE Altai

Period (cal. years BP)			Number of samples					
			Sample location*					Total
			1	2	3	4	5	
Aktru stage (LIA)	100	700	33	4	2			39
	700	1200	10	1		5		16
	1200	1700	2	2	1	1		6
Historical stage	1700	2300						0
	2300	2900		3	2			5
	2900	3500	4	2	2	1		9
	3500	4200	3	1	3			7
Akkem stage	4200	4900	1	1	2			4
	4900	5600	9	4				13
	5600	6400	4	3	1			8
	6400	10000	4	1			1	6
			70	22	13	7	1	113

* Numbers indicate sample locations in 1- North Chuya range; 2- Katun range; 3- South Chuya range; 4- Kurai range; 5- Chikhachev range.

Thus to estimate glacial erosion rate within high mountain area of the Chon-Kyzylsu river basin (Tien Shan, Central Asia) the sediment yield of rivers with glacier alimentation was measured (Iveronova, 1969). The calculated rate of sediment evacuation by glaciers was $1.278 \cdot 10^{-3}$ m a^{-1} (Iveronova, 1969) which is close to presented results of our numerical simulation. It should be mentioned that applied in that work methodic based on measuring the volume of sediments transported by rivers has its drawbacks and limitations. On the one hand the total amount of transported sediments incudes debris coming from valley slopes, on the other - the moraine material, which is accumulated within the glacier valley, is not taken into consideration.

Measuring the total volume of glacial deposits, evacuated from the territory of glacial basins of a certain area during some time period, allows calculating the *effective rate* of *glacial erosion* (Hallet et al., 1996). This rate differs from the *absolute rate of glacial erosion*. These two rates are connected to each other with the simple equation:

$$\dot{h}_{ef} = \frac{S_G}{S} \dot{h}_{ab}.$$

Here $\dot{h}_{ef}$ and $\dot{h}_{ab}$ are effective and absolute rates of glacial erosion correspondingly; S_G - an area of glaciation within the study basin; S - a basin area.

The disadvantages of this technique include problems of determining the glacial deposits among deposits of other origin within sediments of periglacial basins, as well as estimating a volume of deposits and an area of glaciation, which is changing in time. According to (Hallet et al., 1996) the minimal calculation error in the first case is about 10-50%, and in the second one – 10-20%. Calculated in this way the effective rate of glacial erosion will be less than absolute one. Thus for Raikot Glacier (Nanga Parbat Massif, Himalaya) the first rate is (1.4-2.1)10^{-3} m a^{-1} (Table 4), when the second one is (4.6-6.9)10^{-3} m a^{-1}.

Table 4 presents numerical estimates of glacial erosion rate for some Central Asia glaciers, which were obtained by different techniques. Central Asia collision belt is characterized by high seismic activity. Despite of continental climate and ongoing global warming the high absolute altitudes of the Tien Shan, Pamir and Himalaya mountain systems determine the development of powerful modern glaciation there. Altitudes of the Altai mountain ridges are significantly lower, and mountain glaciers in this part of Central Asia are "miniature copies" of glaciers from those grand mountain systems.

One of the largest presented erosion rates – $4.36 \cdot 10^{-3}$ m a^{-1}, refers to the Fedchenko mountain-valley glacier, Pamir, which is characterized by the largest thickness and length. At the same time even powerful glaciers that are developed on the flattened watersheds with the gentle inclinations, demonstrate erosion rate an order of magnitude lower. Thus in Barguzin range, North Baikal region, erosion rate of immobile late Pleistocene glaciers of fabric type with the depth about 800 m is only $0.16 \cdot 10^{-3}$ m a^{-1} (Osipov et al., 2003). This could be the reason of sometimes good preservation of unconsolidated Neogene - Early Pleistocene sediments in such locations (Agatova, 2005).

Table 4. Erosion rate of some valley glaciers for different mountain provinces of the Central Asia collision belt

Location	Area, km^2	Length, km	Erosion rate, 10^{-3} m a^{-1}	Notes	References
Modeled glacier with parameters of the SE Altai glaciers		5	1	Absolute rate Numerical simulation	Agatova and Nepop, 2010
Maly Aktru glacier SE Altai	3.5	3	0.13–0.15	Absolute rate Measurement of sediment evacuation by glaciers	Dushkin, 1974
Zeravshan glacier Hissar-Alay	134	27.8	4.98	Effective rate Measuring sediment yield from glaciers	Chernova, 1981
Imat glacier, Pamir-Alay	3.8		0.83		
Fedchenko glacier Pamir	662	77	4.36		
RGO glacier Pamir	109	24.2	3.19		
Ajutor glacier Tien Shan	3.4		0.24		
Karabulak glacier Tien Shan	4.7		0.38		
Glaciers of the Chon-Kyzylsu basin Tien Shan	4		1.278	Absolute rate Measurement of sediment yield from glaciers	Iveronova, 1969
Raikot Glacier Nanga Parbat, Himalaya	56	14	4.6–6.9	Absolute rate Measurement of sediment yield from glaciers	Gardner and Jones, 1993
Basin of the Raikot Glacier Nanga Parbat, Himalaya	93.3		1.4–2.1	Effective rate Measurement of sediment yield from glaciers	
Siachen Glacier Karakoram, Himalaya	700	70	0.252	Absolute rate Measurement of sediment yield from glaciers	Bhuyiyani, 2000
Gangotri Glaciar Himalaya	258.6	30.2	1.8		Haritashya et al., 2006
Dokriani Glacier Himalaya	7	5.5	1		Singh et al., 2003

Findings of unconsolidated sediments and weathering crusts directly under the retreated glaciers or within the areas of their recent spreading most likely evidence for irregular distribution of glacial erosion and local topography inequalities than for minor role of erosion and exaration activity of moving ice (as it was supposed in (Chistjakov and Seliverstov, 1999, Chistjakov et al., 2012)). Along with the areas of accelerated erosion there are areas with reduced erosion. For example, glacial erosion rates in cirques are an order of magnitude lower than in mountain valleys affected by glaciation (Dushkin, 1974). One of the classic experiments revealed that 3 mm in depth X-shaped label scratched in rocks in front of Dachstein glacier, Austria, was completely erased during its brief advance and the rock surface was polished and striated (Rice, 1988). Similar experiments were repeatedly carried out and always indicated high rates of glacial erosion.

Numerous field observations (Hodge, 1974; Kamb et al., 1979; Paterson, 1994; and others) indicate ice thickness and the rate of its flow to be main factors that control glacial erosion. Physical experiments on modeling of glacial erosion (Khruschev and Babicheva, 1970; Budd et al., 1979; and others) confirmed a significant intensification of bedrock abrasion with increasing normal pressure determined by ice thickness. Erosion activity of fast-moving glaciers can be several times greater than this value for inactive glaciers. Thickness of the basal layer, which is enriched by moraine material, in ice flow is significantly greater than that in inactive part of glacier (Evteev, 1964). Thus more powerful valley glaciers with the maximal height difference of the top and terminal points demonstrate the higher erosion rates in comparison with glaciers located at the flattened watersheds with smoothed topography.

We have also analyzed evolution of glacier bedrock profile for qualitative assessment of glacial effects on topography changes. Numerical simulation demonstrates that under stable climate conditions a significant deepening of the valley occurs at a time interval of about 20 ka (modeling period) (Figure 8).

For clearness erosion rate in the second experiment was increased five times which is equivalent to significantly more powerful glaciations. Results of numerical simulation are presented in Figure 9 where bedrock profiles are shown with the time lag 4 ka. As it was discussed above, taking into account all simplifications and assumptions calculated longitudinal profiles are smoother and less deep than profiles produced by real glaciers. However, this experiment suggests the following conclusion: despite the fact that modern glaciers most likely couldn't affect significantly the bedrock topography even during the next 10 ka, erosion activity of Pleistocene glaciers (which were much more powerful) caused serious deepening of mountain valleys. Erosion activity of mountain glaciers is not limited only by widening of the valleys and giving them a U-shaped cross-section profile (as it is suggested in (Chistjakov et al., 2000)). The same way as rivers act during interglacial periods, glaciers deepen the valleys during glacial periods. Glaciers sizes and duration of their influence are the main factors that control glacial erosion. Analyzing the heights of troughs shoulders and estimating the Pleistocene erosion rate for Maly Aktru glacier, Dushkin (1974) concluded the glacial origin of mountain valleys deepening within the SE Altai. The same conclusion was made as a result of numerical simulation of glaciation in the Yosemite Valley, Sierra Nevada Mountains, Northern California (MacGregor et al., 2000).

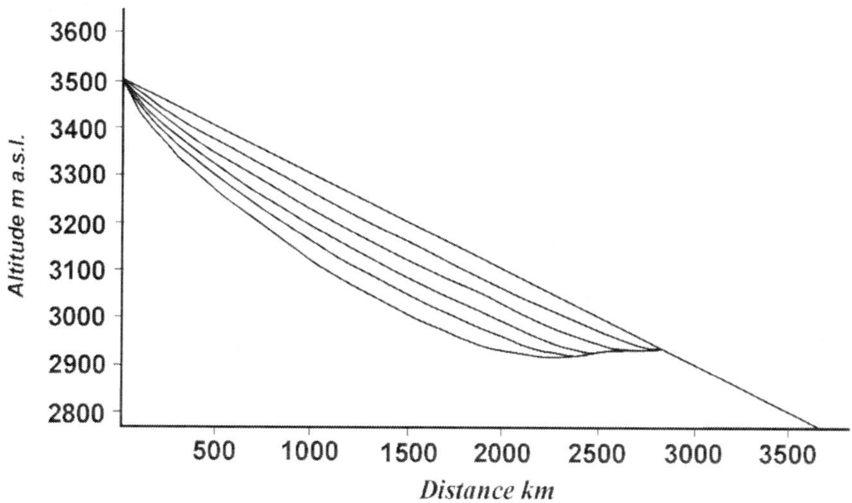

Figure 9. 20 ka evolution of glacial valley longitudinal profile. Profiles are shown with the 4 ka interval.

According to our simulation results the largest volume of rock material is removed from the bottom of the valley near the ELA position. Particularly this part of glacier bed is most of all affected by glacial erosion and is characterized by the highest erosion rate. Numerical model more correctly describes the behavior of real glaciers in the upper and middle parts of the valley. So these parts of modeled longitudinal profiles are most closely corresponds to the real ones. The low part of real glacial valleys is complicated by accumulation of moraine material and erosion activity of ice-melted water.

Accelerating erosion rate is observed just before the accumulation zone in the low part of the real glacial valley. It causes local overdeepening of the longitudinal profile (Alley et al., 2003). Modeled profiles also demonstrate formation of such overdeepenings. Local decrease of bedrock inclination in the low part of the valley leads to increasing ice thickness.

Figure 10 demonstrates glacier transition to a stationary state. Under the permanent climate conditions implied in the numerical simulation the model glacier practically ceases changing its shape and enters into a stationary state at the time interval of about 10^2 year. Since that time, appearance of a glacial valley is defined by internal processes of mass transfer in a glacier and glacial erosion on bedrock. This experiment was interesting to estimating the length response time which is typical for some SE Altai glaciers. This parameter characterizes the reaction time of glacier in a stationary state to a sudden massbalance change (Paterson, 1994). Length response time calculated for Sofiysky Glacier (SE Altai) is 73-84 years, for Maly Akrtu Glacier (SE Altai) - 99-114 years (De Smedt and Pattyn, 2003), and for Hintereisferner Glacier (Austrian Alps), which is characterized by the similar topographic features, is 94 ± 15 years (Greuell, 1992). Thus, presented numerical model quite correctly describes some dynamic parameters of real glaciers.

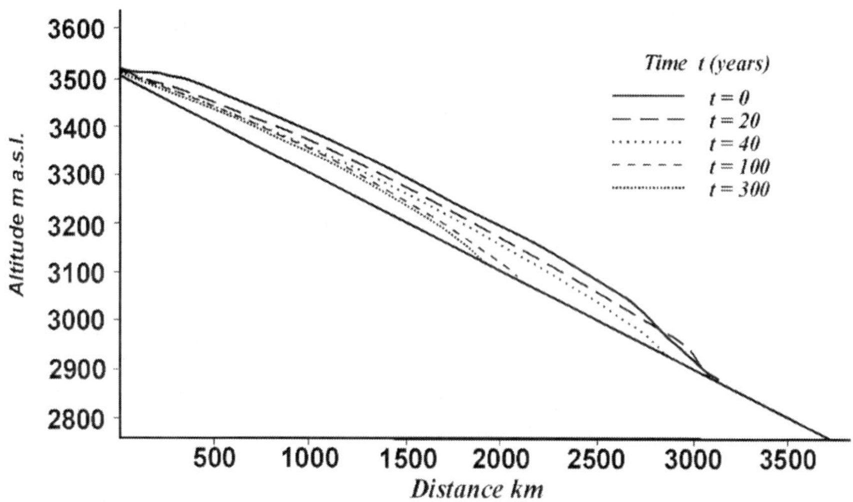

Figure 10. The reaction of the steady state glacier (at $t = 0$) to a stepwise change in a glacier mass balance.

Obtained results also demonstrate that the SE Altai glaciers can serve as reliable markers of only long-periodic climate variations (with a period of 100 years and more). Numerous field observations and theoretical calculations demonstrate sensitive reaction of glacier tongue to the sudden massbalance change (Paterson, 1994). However, applying data on high-frequency oscillations of the glacier front (with the period less than length response time) for climatic reconstructions is quite difficult due to the problems of taking into consideration the entire previous history of glacier development. At the present time it is practically impossible to determine reliably the superposition of what particular climatic events affected changes in glacier length and ice thickness.

CONCLUSION

Studying glacial and associated landforms in the SE Altai, Russia, allows specifying the Holocene chronology of glaciers advances within one of the most northern segments of the Central Asia Collision Belt. Numerous radiocarbon ages of buried in moraines and located above the modern upper timber limit wood fragments was used to establish the time periods favorable for forest growth in modern glaciated areas and to identify the upper tree limit position associated with prolonged and steady climatic improvements. The extended data set of such favorable periods limits the times of climate deterioration and possible glacier expansion.

The collective evidences indicate that the early Holocene climate was warmer than modern one. There were no traces of the late Würm (Sartan) glaciation not only within intermountain depressions and at the foots of framing ridges, but also at the heads of troughs already in the beginning of the Holocene and mountain glaciers were either completely decayed or reached only the modern extent. By now there are still not enough absolute age determinations for specifying the climatically driven glaciers dynamics in the first half of the Holocene. The second half of the Holocene, Neoglacial, is characterized by shorter glacier recessions and more prolonged and frequent advances compared to the first half.

Three clearly expressed in topography moraine complexes correlate with glaciers advances during the Akkem stage (about 4900-4200 cal. years BP), Historical stage (2300-1700 cal. years BP), and LIA or Aktru stage (the 13th - 19th centuries). Successively less Holocene glacier expansion in each subsequent stage could be driven by progressive aridization which was intensified in this part of Central Asia during the last 2 ka. During the periods between expansions glaciers degraded to at least the modern state, and possibly even more. These degradations were accompanied by repeated forest regrowth in the presently glaciated area above the modern upper tree limit.

Numerical simulation of the ice movement in the mountain-valley glacier and its interaction with the valley bottom was applied for calculating glacial erosion rate and length response time for the SE Altai glaciers as well as for qualitative analysis of the bedrock topography evolution at the time interval of about 10^3–10^5 years. The presented one-dimensional numerical model of idealized glacier describes the ice flow along a central flow line. Physical constants, hypsometric data and other parameters of model glacier are closely corresponded to modern climatic, geological and geomorphological characteristics of the SE Altai. Calculated absolute glacier erosion rate is about $1.0 \cdot 10^{-3}$ m a^{-1} with the peak value $2.0 \cdot 10^{-3}$ m a^{-1} near ELA. The reaction time to a stepwise change of the surface mass balance was calculated as about 10^2 years. These estimates are in agreement with those obtained by other researchers of modern glaciers in other mountain systems within the Central Asian Collision Belt.

In such mountain systems as Altai, where modern glaciation is preserved only within axial parts of the highest ridges, glacial erosion affects only the upper tier of mountain topography. Repeated late Pleistocene and Holocene glaciers advances leaded to expanding the area affected by glaciers and increasing glacial erosion rate. Influence of Quaternary glaciations on topography evolution was not limited by formation of typical U-shaped cross-section profiles, but was also resulted in significant deepening of glacial valleys.

REFERENCES

Adamenko, M.F.; Selishchev, E.N. New data on glacier dynamics in the Akturu and Kurumdu river basins during the Little Ice Age. In *Nature and economics of the Kuznetsk Basin;* Novokuznetsk, 1984; pp.58-61. (In Russian).

Adamenko, M.F.; Syubaev, A.A. Climate dynamics in the territory of mountain Altai in the 15th–20th centuries on the basis of dendrochronological analysis. In *Problems of Mountain Glaciology*; Tomsk University Press: Tomsk, 1977; pp. 196-202. (In Russian).

Agatova, A.R. Geomorphologic mapping of the Chagan-Uzun river basin: a key for reconstructing history of pleistocene glaciations in the Southeastern Altai. *Stratigraphy and Geological Correlation.* 2005, 13(6), 656-666.

Agatova, A.R.; Nepop, R.K. Glacier erosion rate, response time of modern glaciers and evolution of longitudinal profile of glacial valleys of south-eastern Altai on the basis of numerical simulation. *Ice and Snow*. 2010, 4, 111-120. (In Russian).

Agatova, A.R.; Van Huele, W.; Mistrukov, A. Dynamics of Sofiysky glacier (SE Altai): Last Glacial - 20th century. *Geomorfologiya*. 2002, 2, 92-104. (In Russian).

Agatova, A.R.; Nazarov, A.N.; Nepop, R.K.; Rodnight, H. Holocene glacier fluctuations and climate changes in the southeastern part of the Russian Altai (South Siberia) based on a radiocarbon chronology. *Quaternary Science Reviews*. 2012, 43, 74-93.

Agatova, A.R.; Nepop, R.K.; Barinov, V.V.; Nazarov, A.N.; Myglan, V.S. The first dating of strong Holocene earthquakes in Gorny Altai using long-term tree-ring chronologies. *Russian Geology and Geophysics*. 2014a, 55(9), 1065-1073.

Agatova, A.R.; Nepop, R.K.; Slyusarenko, I.Yu.; Myglan, V.S.; Nazarov, A.N.; Barinov, V.V. Glacier dynamics, palaeohydrological changes and seismicity in southeastern Altai (Russia) and their influence on human occupation during the last 3000 years. *Quaternary International*. 2014b, 324, 6-19.

Albrecht, O.; Jansson, P.; Blatter, H. Modelling glacier response to measured mass balance forcing. *Annals of Glaciology*. 2000, 31, 91-96.

Alley, R.B.; Lawson, D.E.; Larson, G.J.; Evenson, E.B.; Baker, G.S. Stabilizing feedbacks in glacier-bed erosion. *Nature*. 2003, 424, 758-760.

Bhuyiyani, M.R. Sediment load characteristics of a proglacial stream of Siachen Glacier and the erosion rate in Nubra valley in the Karakoram Himalayas, India. *Journal of Hydrology*. 2000, 227, 84-92.

Blyakharchuk, T.; Wright, H.; Borodavko, P.; Knaap, W.O. van der; Ammann, B. The role of pingos in the development of the Dzhangyskol lake-pingo complex, central Altai Mountains, southern Siberia. *Palaeogeography, Palaeoclimatology, Palaeoecology*. 2008, 257, 404-420.

Budd, W.F.; Keage, P.L.; Blundy, N.A. Empirical studies of ice sliding. *Journal of Glaciology*. 1979, 23, 157-170.

Bulatov, V.I.; Dik, I.P.; Revjakin, V.S. Glaciological observations in Akkol-river basin. *Glyatsiologiya Altaya*. 1967, *5*, 178-183 (in Russian).

Caidong, C.; Sorteberg, A. Modeled mass balance of Xibu glacier, Tibetan Plateau: sensitivity to climate change. *Journal of Glaciology*. 2010, 56(196), 235-248.

Catalogue of glaciers of the USSR. Chuya basin. Part 6. GIMIZ, Leningrad, 1974, 15(1). (In Russian).

Chernova, L.P. Intensity of glacial erosion and its relation to the ice discharge in glaciers. *Geomorfologija*. 1974, 4, 12-18. (In Russian).

Chernova, L.P. Influence of mass balance and run-off on relief-forming activity of mountain glaciers. *Annals of Glaciology*. 1981, 2, 69-70.

Chistjakov, K.V.; Seliverstov, Yu.P. *Regional ecology of little changed landscapes: North-East of Inner Asia;* Publishing house of SPb University: SPb., 1999; pp. 264. (In Russian).

Chistjakov, K.V.; Ganjushkin, D.A.; Moskalenko, I.G.; Zelepukina, E.S.; Amosov, M.I.; Volkov, I.V.; Glebova, A.B.; Guzel', N.I.; Zhuravlev, S.A.; Prudnikova, T.N.; Prjahina, G.V. *Mountain massif Mongun-Taiga;* Art-Express: St. Petersburg, 2012; pp. 310. (in Russian).

Chistjakov, A.A.; Makarova, N.V.; Makarov, V.I. *Quaternary geology*; GEOS: Moscow, 2000; pp. 303. (In Russian).

De Smedt, B.; Pattyn, F. Numerical modeling of historical front variations and dynamic response of Sofiyskiy glacier, Altai mountains, Russia. *Annals of Glaciology*. 2003, 37, 143-149.

Devyatkin, E.V. *Cenozoic sediments and neotectonics of the SE Altai;* USSR Academy of Science: Moscow, (GIN Transactions, Issue 126), 1965; pp. 244. (in Russian).

Dushkin, M.A. Perennial fluctuations of Aktru glaciers and conditions of young moraines. *Glyatsiologiya Altaya*. 1965, 4, 83-101. (In Russian).

Dushkin, M.A. Erosion of glaciers of the North Chuya range, Central Altai. *Gljaciologija Altaja*. 1974, 8, 28-38. (In Russian).

Dyurgerov, M.B.; Meier, M.F. Year-to-year fluctuations of global mass balance of small glaciers and their contribution to sea-level changes. *Arctic, Antarctic and Alpine Research.* 1997, 29(4), 392-402.

Evteev, S.A. *Geological activity of ice sheets in Eastern Antarctic;* Nauka: Moscow, 1964; pp. 120. (In Russian).

Fujita, K.; Ageta, Y. Effect of summer accumulation on glacier mass balance on the Tibetan Plateau revealed by mass-balance model. *Journal of Glaciology.* 2000, 46(153), 244-252.

Galahov, V.P.; Nazarov, A.N.; Harlamova, A.N. *Glaciers fluctuations and climate changes during late Holocene on the basis of studying of glaciers and glacial sediments of Aktru basin (Central Altai, North Chuya range);* Altai University Press: Barnaul, 2005; pp. 132. (In Russian).

Galahov, V.P.; Nazarov, A.N.; Lovskaya, O.V.; Agatova, A.R. *Chronology of the warm period in the second half of the Holocene in SE Altai (on the basis of radiocarbon dating of glacial sediments);* Azbuka Publisher: Barnaul, 2008; pp. 58. (In Russian).

Galahov, V.P.; Muhametov, R.M. *Altai glaciers;* Nauka: Novosibirsk, 1999; pp. 136. (In Russian).

Galahov, V.P.; Chernyh, D.V.; Zolotov, D.V.; Agatova, A.R.; Birjukov, R.Yu.; Orlova L.A.; Ostanin, O.V.; Samojlova, S.Yu.; Sheremetov, R.T.; Yakubovsky, V.I. *Glaciation of the southwestern part of Altai in the second half of the Holocene;* Azbuka Publisher:Barnaul, 2012; pp. 110. (In Russian).

Gardner, J.S., Jones N.K. Sediment transport and yield at the Raikot Glacier, Nanga Parbat, Punjab Pakistan. In: *Himalaya to the Sea: Geology, Geomorphology at the Quaternary;* Routledge: London, 1993; pp. 184-197.

Greuell, W. Hintereisferner, Austria: mass balance reconstruction and numerical modeling of the historical length variation. *Journal of Glaciology*. 1992, 38(129), 244-252.

Haeberli, W., Hoelzle, M. Application of inventory data for estimating characteristics of and regional climate-change effects on mountain glaciers: a pilot study with the European Alps. *Annals of Glaciology.* 1995, 21, 206-212.

Hallet, B. A theoretical model of glacier abrasion. *Journal of Glaciology.* 1979, 23, 39-50.

Hallet, B. Glacial quarrying: A simple theoretical model. *Annals of Glaciology*. 1996, 22, 1-8.

Hallet, B.; Hunter, L.; Bogen, J. Rates of erosion and sediment evacuation by glaciers: A review of field data and their implications. *Global and Planetary Change*. 1996, 12, 213-235.

Haritashya, U.; Singh, P.; Kumar, N.; Gupta, R.P. Suspended sediment from the Gangotri Glaciar: Quantification, variability and associations with discharge and air temperature. *Journal of Hydrology*. 2006, 321, 116-130.

Hodge, S.M. Variations in the sliding of a temperate glacier. *Journal of Glaciology*. 1974, 13, 349-369.

Hooke, R.L. Positive feedbacks associated with erosion of glacial cirques and overdeepenings. *Geological Society of American Bulleting*. 1991, 103, 1104-1108.

Humphrey, N.F.; Raymond, C.F. Hydrology, erosion and sediment production in a surging glacier: Variegated Glacier, Alaska, 1982–83. *Journal of Glaciology*. 1994, 40, 539-552.

Huybrechts, Ph.; De Nooze, P.; Decleir H. Numerical modeling of glacier D'Argentiere and its historic front variations. In: *Glacier Fluctuations and Climatic Change;* Kluwer Academic Publisher, 1989; pp. 373-389.

Ivanovsky, L.N. *Glacial landforms and their palaeogeographical importance in the Altai*; Nauka: Leningrad, 1967; pp. 234. (In Russian).

Ivanovsky, L.N. *Exogenous litho dynamics of mountain provinces;* Nauka: Novosibirsk, 1993; pp. 160. (In Russian).

Ivanovsky, L.N.; Panychev, V.A. The development and age of terminal moraines of the XVII-XIXth centuries of Aktru glaciers, Altai. In *Present Relief Formation Processes in Siberia*; Nauka: Irkutsk, 1978; pp.127-138. (In Russian).

Ivanovsky, L.N.; Panychev, V.A.; Orlova, L.A. Age of terminal moraines of "Aktru" and "Historical" stages of Altai glaciers. In *Late Pleistocene and Holocene of the southern part of Eastern Siberia;* Nauka: Novosibirsk, 1982; pp. 57-64. (In Russian).

Iveronova, M.I. An experience of quantitative analysis of modern denudation. *Izvestiya AN SSSR, Ser. Geographical.* 1969, 2, 13-24. (In Russian).

Johnsen, S.J.; Clausen, H.B.; Dansgaard, W.; Gundestrup, N.S.; Hammer, C.U.; Andersen, U.; Andersen, K.K.; Hvidberg, C.S.; DahlJensen, D.; Steffensen, J.P.; Shoji, H.; Sveinbjornsdottir, A.E.; White, J.; Jouzel, J.; Fisher, D. The delta O-18 record along the Greenland Ice Core Project deep ice core and the problem of possible Eemian climatic instability. *Journal of Geophysical Research-Oceans.* 1997, 102, 26397-26410.

Kamb, B.; Engelhardt, H.F.; Harrison, W.D. The ice-rock interface and basal sliding processes as revealed by direct observation in bore holes and tunnels. *Journal of Glaciology*. 1979, 23, 416-419.

Khruschev, M.M.; Babicheva, M.A. *Abrasive wearing;* Nauka: Moscow, 1970; pp. 252. (In Russian).

Lehmkuhl, F.; Zander, A.; Frechen, M. Luminescence chronology of fluvial and aeolian deposits in the Russian Altai (Southern Siberia). *Quaternary Geochronology.* 2007, 2, 195-201.

Le Roy, M.; Nicolussi, N.; Deline, P.; Astrade, L.; Edouard, J-L; Miramont, C.; Arnaud, F. Calendar-dated glacier variations in the western European Alps during the Neoglacial: the Mer de Glace record, Mont Blanc massif. *Quaternary Science Reviews.* 2015, 108, 1-22.

MacGregor, K.R.; Anderson, R.S.; Anderson, S.P.; Waddington, E.D. Numerical simulations of glacial-valley longitudinal profile evolution. *Geology*. 2000, 28(11), 1031-1034.

Mayewski, P.A.; Rohling, E.E.; Curt Stager, J.; Karlen, W.; Maasch, K.A.; David Meeker, L.; Meyerson, E.A.; Gasse, F.; van Kreveld, S.; Holmgren, K.U. Holocene climate variability. *Quaternary Research.* 2004, 62, 243-55.

Mikhailov, N.N. The dynamics of Belukha glaciers, Altai, in the historical time. *Bulletin of Leningrad State University. Geology-Geography Series.* 1987, 3, 100-103. (In Russian).

Mikhailov, N.N.; Maksimov, E.V.; Kozyreva, M.G.; Larin, S.I.; Merkulov, P.I.; Chernov, S.B. Radiocarbon dating of the Holocene sediments from mountain provinces of the southern framing of the USSR. *Bulletin of Leningrad State University. Geology-Geography Series.* 1989, 1(7), 57-62. (In Russian).

Mjagkov, I.M. Moraines of Beluha glaciers. *Bulletin of the Western Siberia Geological Trust.* 1936, 1, 85-106. (In Russian).

Myglan, V.S.; Ojdupaa, O.Ch.; Vaganov, E.A. Making the 2367-year tree-ring chronology for Altai-Sayan region (mountain massif Mongun-Taiga). *Archeology, etnography and anthropology of Eurasia.* 2012, 3, 76-83. (in Russian).

Naroznyi, Yu.K.; Osipov, A.V. Oroclimatic conditions of the Central Altai glaciations. *Izvestiya RGO.* 1999, 131(3), 49-57. (In Russian).

Nazarov, A.N.; Agatova, A.R. Glacier dynamics in the North Chuya range, Central Altai, during the second half of the Holocene. *Materials of Glaciological Investigations.* 2008, 105, 73-86. (in Russian).

Nazarov, A.N.; Myglan, V.S. The possibility of construction of the 6000-year chronology on Pinus Sibirica in the Central Altai. *Journal of Siberian Federal University.* 2012, 1, 70-88. (in Russian).

Nazarov, A.N.; Solomina, O.N.; Myglan, V.S. Variations of the Tree Line and Glaciers in the Central and Eastern Altai Regions in the Holocene. *Doklady Earth Sciences.* 2012, 444, 787-791.

Nepop, R.K. Numerical assessing the Holocene rates of exogenous relief-making processes within the Chagan-Uzun River basin. *Abstracts of the PhD Thesis;* Institute of Earth Crust SB RAS: Irkutsk, 2007; pp. 16. (in Russian).

Nikitin, S.A.; Osipov, A.V.; Vesnin, A.V.; Iglovskaya, N.V. Distribution of fresh water resources in Central Altai glaciers. In *Fundamental problems of water and water resources at the edge of the third millennium*; NTL Press, Tomsk, 2000, pp. 341-345. (In Russian).

Nikitin, S.A.; Tatarinov V.N. Applying radio-echo sounding for studying Altai glaciers. *Materials of Glaciological Investigations*. 1982, 44, 156-164. (in Russian).

Nye, J.F. The effect of longitudinal stress on the shear stress at the base on ice sheet. *Journal of Glaciology*. 1969, 8, 207-213.

Oerlemans, J.A. Flowline model for Nigardsbreen, Norway: Projection of future glacier length based on dynamic calibration with the historic record. *Annals of Glaciology*. 1997, 24, 382-389.

Okishev, P.A. *The dynamics of glaciation in Altai during the Late Pleistocene and Holocene;* Tomsk University Press: Tomsk, 1982; pp. 210. (In Russian).

Okishev, P.A. *Topography and glaciations of Rusian Altai;* Tomsk University Press: Tomsk, 2011; pp. 382. (in Russian).

Osipov, E.Yu.; Grachev, M.A.; Matc, V.D.; Khlystov, O.M.; Braitenbah S. Reconstruction of mountain glaciers of the last Pleistocene glaciation in the northwestern part of Barguzin range (North Baikal region). *Geology and geophysics*. 2003, 7(44), 652-663. (In Russian).

Pattyn, F.; De Smedt, B.; Van Huele, W.; Agatova, A.; Mistrukov, A.; Decleir, H. Ice dynamics and basal properties of Sofiyskiy Glacier, Altai Mountains, Russia, based on DGPS and radio-echo sounding surveys. *Annals of Glaciology*. 2003, 37, 286-292.

Paterson,W.S.B. *The physics of glaciers*; Pergamon Press: Oxford, New York, Tokyo, 3rd Edition, 1994; pp. 480.

Reimer, P.; Baillie, M.; Bard, E.; Bayliss, A.; Beck, J.; Blackwell, P.; Bronk; Ramsey, C.; Buck, C.; Burr, G.; Edwards, R.; Friedrich, M.; Grootes, P.; Guilderson, T.; Hajdas, I.; Heaton, T.; Hogg, A.; Hughen, K.; Kaiser, K.; Kromer, B.; McCormac, F.; Manning, S.; Reimer, R.; Richards, D.; Southon, J.; Talamo, S.; Turney, C.; Plicht, J.; Weyhenmeyer, C. IntCal09 and Marine09 radiocarbon age calibration curves, 0-50,000 years cal BP. *Radiocarbon*. 2009, 51(4), 1111-1150.

Revjakin, V.S.; Galahov, V.P.; Goleschihin, V.P. *Mountain glacier basins of Altai;* Tomsk University Press: Tomsk, 1979; pp. 309. (In Russian).

Rice, R.J. *Fundamentals of Geomorphology;* Longman: London, 1988; pp. 420 p.

Rusanov, V.I. Distribution of annual precipitation in Central Altai. *Russian Geological Society Proceedings*. 1961, 93(6), 272-283. (In Russian).

Sapozhnikov, V.V. *The ways along the Russian Altai;* Tomsk, 1912, pp. 170. (In Russian).

Schlütz, F.; Lehmkuhl, F. Climatic change in the Russian Altai, southern Siberia, based on palynological and geomorphological results, with implications for climatic teleconnections and human history since the middle Holocene. *Vegetation History and Archaeobotany*. 2007, 16, 101-118.

Shnitnikov, A.V. *Variability of general moistening of the continents in the Northern Hemisphere*; Reports from the USSR Geographical Society; Nauka: Moscow-Leningrad, 1957; Vol.16, pp. 338. (In Russian).

Singh, P.; Ramasastri, K.S.; Kumar, N.; Bhatnagar, N.K. Suspended sediments transport from the Dokriani Glacier in the Garhwal Himalayas. *Nordic Hydrology*. 2003, 34, 221-244.

Slusarenko, I.Yu. *Dendrochronological dating of the archeological monuments of Scythian epoch of Altai. Abstracts of the PhD Thesis;* IAiE: Novosibirsk, 2010; pp. 34. (In Russian).

Slusarenko, I.Yu.; Christen, J.A.; Orlova, L.A.; Kuzmin, Y.V.; Burr, G.S. C-14 Wiggle matching of the "floating" tree-ring chronology from the Altai Mountains, Southern Siberia: the Ulandryk 4case study. *Radiocarbon*. 2001, 43(2A), 417-424.

Solomina, O.N. The lichenometry of moraines at the Bish-Iirdu mountain massif. *Materials of Glaciological Investigations;* 1988, 61, 118-123. (In Russian).

Solomina, O.N. *Mountain glaciations of the Northern Eurasia during the Holocene*; Nauchny Mir: Moscow, 1999; pp. 272. (In Russian).

Solomina, O.; Bradley, R.; Hodgson, D.; Ivy-Ochs, S.; Jomelli, V.; Mackintosh, A.; Nesje, A.; Owen, L.; Wanner, H.; Wiles, G.; Young, N. Holocene glacier fluctuations. *Quaternary Science Reviews*. 2015, 111(1), 9-34.

Stuiver, M.; Reimer, P.J. Extended C-14 data-base and revised calib 3.0 C-14 age calibration program. *Radiocarbon*. 1993, 35, 215-230.

Tjumencev, K.G. *The report on geology-glaciological part of the Altai glacial expedition of the year 1933*; Transactions of glacial expedition of the USSR Academy of Sciences; USSR AS publishers: Moscow, 1936; Issue 6, pp. 37-94. (In Russian).

Vardanyants, L.A. About ancient glaciations of the Altai and the Caucasus (comparative sketch). *Izvestiya USSR geographical society*. 1938, 70(3), 386-406. (in Russian).

Zol'nikov, I.D.; Deev, E.V.; Lyamina, V.A. New data on Quaternary morpholithogenesis in the Chuya basin (Gorny Altai). *Russian Geology and Geophysics*. 2010, 51(4), 339-348.

Biographical Sketch

Roman Nepop

Affiliations:

Institute of Geology and Mineralogy SB RAS, Novosibirsk, Russia
and Ural Federal University, Yekaterinburg, Russia

Education:

Novosibirsk State University, Muster of science in physics, PhD in geology.

Address: Koptyuga av., 3, Novosibirsk, Russia, 630090.

Research and Professional Experience:

Research and professional experience includes geological and geomorphological investigations based on various scientific methods and geochronological techniques in high mountain provinces of Central Asia: Russian and Mongolian Altai, SW Tuva, SW Mongolia. Participant of numerous expeditions in Russian and Mongolian Altai, Eastern Sayan, Tien Shan, Mongolia, Polar Yakutia.

Studying paleolandscape evolution, regional hydrological system transformation related to the evolution of giant Pleistocene ice-dammed and the Holocene residual lakes, climatically driven Holocene glacier dynamics, paleoseismogeological investigations including dendroseismological approach, studying the possibility of absolute dating of glacial deposits and landforms, geoarchaeological researches fall in the area of current interests.

Professional Appointments: Senior researcher

Publications Last 3 Years:

Agatova, A.R., Nepop R.K., Slyusarenko I.Yu., Myglan V.S., Nazarov A.N., Barinov V.V. Glacier dynamics, palaeohydrological changes and seismicity in southeastern Altai (Russia) and their influence on human occupation during the last 3000 years. *Quaternary International*. 2014. V.324. P.6-19

Agatova, A.R., Nepop, R.K., Barinov, V.V., Nazarov, A.N., Myglan, V.S. A new aspect of application of dendrochronological analysis for dating strong earthquakes of the past: A case study of the Altai Mountains. *Doklady Earth Sciences*. 2014. V.455(1). P.243-245.

Nepop, R.K., Agatova, A.R., Myglan, V.S., Nazarov, A.N., Barinov, V.V. Using dendrochronological analysis for dating earthquake-triggered landslides (by the example of SE Altai, Russia). In: *Landslides Science for a Safer Geoenvironment*. Sassa K., Canuti P., Yin Y. (Eds.) 2014. Springer. P. 257-263.

Agatova, A.R., Nepop, R.K., Bronnikova, M.A., Slyusarenko, I.Yu., Orlova, L.A. Human occupation of South Eastern Altai highlands (Russia) in the context of environmental changes. *Archaeological and Anthropological Sciences*. 2014. DOI 10.1007/s12520-014-0202-7

Agatova, A.R., Nepop, R.K., Barinov, V.V., Nazarov, A.N., Myglan, V.S. The first dating of strong Holocene earthquakes in Gorny Altai using long-term tree-ring chronologies. *Russian Geology and Geophysics*. 2014. V.55. P.1059–1067.

Tainik, A.V., Myglan, V.S., Barinov, V.V., Nazarov, A.N., Agatova, A.R., Nepop, R.K. Tree-Ring Growth of Siberian Larch at the Upper Treeline in the Altai Republic. Izvestiya RAN. *Seriya Geograficheskaya*. 2015. V.6. P.75-85. (In Russian).

Baryshnikov G., Agatova A., Carling P., Herget J., Panin A., Adamiec G., Nepop R. Russian Altai in the Late Pleistocene and the Holocene: Geomorphological catastrophes and landscape rebound. Barnaul: Publishing House of Altai State University. 2015. 137 p.

Agatova, A.R., Nepop R.K., Bronnikova M.A. Outburst floods of the ice-dammed lakes in the SW of Tuva, southern Siberia. *Zeitschrift für Geomorphologie* (Annals of Geomorphology). 2015 V.59, Suppl.3, P. 159-175.

In: Glaciers
Editor: Nicholas Doyle

ISBN: 978-1-63484-941-8

Chapter 6

DEBRIS COVER EXPANSION ON DJANKUAT GLACIER THROUGH THE DIRECT MONITORING PERIOD AND ITS EFFECT ON THE MASS BALANCE

Alexey Rezepkin*
Department of Cryolithology and Glaciology, Geographical Faculty, Lomonosov Moscow State University, Moscow, Russia

ABSTRACT

During the current stage of deglaciation, occurring almost worldwide, the supraglacial debris cover often encounters attributes of glacier snouts. Moreover, the supraglacial debris cover has a great effect on the ablation, runoff and mass balance of glaciers. Nevertheless, the detailed observation of debris cover had not yet been included in the programs of direct glaciological monitoring. Supraglacial debris cover behavior on the Djankuat glacier (Central Caucasus, Russia), known to be a representative glacier for the Central Caucasus, has been examined in different ways for almost 50 years. Remarkably, the continuity of observation methods and researchers provides comparability of the obtained data. Supraglacial debris has been studied both with direct field techniques and remote cartographical methods. Definitive expansion of debris material with the course of time was found out on the Djankuat

* E-mail: rezepkin@gmail.com.

glacier snout: occupying only 2% of the glacier surface in 1968, the debris-covered part extended up to 13% in 2010 - enlargening, consequently, by 6.5 times over 42 years. From 1983 to 2010, accumulation of debris near the glacier terminus was up to 70 cm, while the volume of debris material through the entire glacier experienced 141% growth. These changes were reflected both in shapes of mass balance fields and absolute values.

INTRODUCTION

Supraglacial debris, covering lower parts of alpine glaciers, are an inherent part of them at the present time. The modern stage of alpine glaciation evolution is generally characterized by active deglaciation, accompanied by intense thickening and spatial expansion of supraglacial debris cover. Supraglacial debris materials change all the processes within the boundaries of glacial areas covered by it, predetermining the ablation rates and runoff from those parts, and controlling the shape of mass-balance fields and ablation fields in particular: thin layer of debris (up to 7 cm) increases subderis ablation, while a thicker layer, by contrast, decreases it up to the full termination (Bozhinskiy et al., 1986).

Having thermal and physical characteristics is different from the clean ice surface; supraglacial debris forms a very specific structure of glacier-atmosphere heat exchange. Therefore, transformation of supraglacial debris morphology leads to change of the whole appearance of the glacier. At the same time, research of supraglacial debris is not included in the standard program of traditional glaciological monitoring. Observation of supraglacial debris expansion on Djankuat glacier (Central Caucasus, Russia) started synchronously with the beginning of glaciological monitoring in 1968 (Boyarskiy 1978). This paper reports a continuation of direct field investigations, aiming to reveal whether the supraglacial debris effect is increasing or decreasing in mass balance and glacier evolution with the course of time.

METHODS

Currently the two main methods of supraglacial debris study on the Djankuat glacier are the direct field investigations and the cartographical observations. Geophysical data is waiting to be interpreted. Moreover, a dynamic high-resolution geodetic survey is scheduled to be conducted in the near future.

The Cartographical method is based on the analysis and comparison of six large scale maps (1:10000) with 10 meter contour intervals, built on the basis of phototheodolite survey of the glacier in 1968, 1974, 1984, 1992, 1996 and 2006, with the first and the last ones covering not only the glacier itself, but the whole glacial basin. All of them provide the information to estimate changes in morphometry through the period of direct monitoring. Their main distinguishing characteristic is conducting the survey from the fixed bedrock bases (Zolotarev and Popovnin 1993). Moreover, all maps have the same false grid system, which have been adjusted to the true geographical grid systems by means of d-GPS only in 2010. Debris-covered areas were marked by the special symbols on these maps.

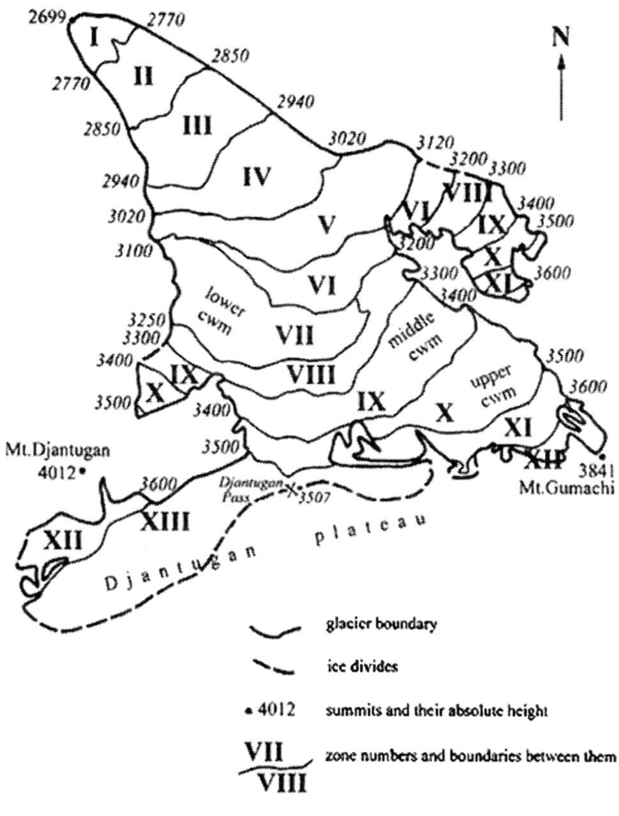

Figure 1. Scheme of height-morphological zones of Djankuat glacier.

All maps were analytically processed through the instrumentality of cartographical computer packages. Furthermore, for in-depth analysis, all glaciological data is traditionally divided into 13 height-morphological zones – parts of the glacier surface, separated by the presence of more or less homogeneous morphological features, with numerical order increasing accordingly with the elevation (Figure 1). The four lower zones annually present the ablation area, zones V and VI could be considered as year after year firn line migration belt, while the upper zones mainly belong to the accumulation area.

However, remote methods turn out to be inapplicable for discovering temporal variations of debris cover thickening. This case requires full-scale direct ground-level debris mapping, including the whole debris-covered part of the glacier. Examples of comprehensive debris mapping on the glaciers of the Earth are singular, and the Djankuat glacier belongs to these rare objects.

The first full investigation of debris thickness on Djankuat glacier for estimating of hydrological effect was conducted in 1983. The procedural framework looked very much like the examination of snow cover: thickness was measured in points spread over the whole debris-covered area. Bassin et al. (1983), while working at Schultz, RGO, Kuldjilga and Podkova glaciers, arranged these points in chess-board fashion, which failed to provide the proper detail level of isolines pictured on knob-and-basin parts of the debris terrain. Profusion of debris mesorelief linear elements, streamlined along the glacier movement axis, combined with huge variations of debris depths in different parts of the glacier, require other solutions. Thus, the landscape principle of debris cover mapping is worthwhile, wherein the chess-board method of probation points spreading on the parts of debris cover with relatively monotonous relief would harmonize with cross profiles of debris ramparts, consisting of points that may be closely spaced to each other, but they would be indicative in terms of the landscape features (bendings, shoulders, peaks, etc.).

Precisely this principle was applied during the debris cover mapping on the Djankuat glacier. "Continuous debris cover" was considered to have projective cover of more than 80%. The measuring procedure included pushing metal battens between the rocks until it reaches the ice surface or hand excavations, which sometimes appeared as full-scale survey pits. Scale readings were taken with accuracy of 1 cm. Natural roughness of the clastic rocky cover in this case was smoothed with the aid of a thin tin ring, which was mounted on the probe and neutralized oscillations of debris roofing. Debris cover thickness was measured in this manner in 133 points, spread over the whole territory with projective cover of supraglacial debris close to 100%. In 1994, this study was repeated with increased accuracy and degree of detail – this time debris cover thickness was measured in 240 points. As a result, two maps of debris cover thickness distribution in the lower part of the Djankuat glacier were obtained for 1983 and 1994, primarily analyzed earlier (Popovnin and Rozova 2002).

The same study was conducted on the Djankuat glacier in the summer of 2010. The measuring method remained the same – as far as possible, hand excavations of debris material for the purpose of visual verification of reaching the surface of ice, buried under the debris; in other cases – penetration of debris stratum. Measurements were made in 189 points throughout the whole debris-covered glacial area. Values of debris cover depths in each point had been manually plotted on the route map, which included the topographic base of the glacier. Moreover, map control was conducted by putting positions of every point into the memory of the GPS-tracker.

Detailed direct field observations of debris cover thickness, unlike determination of this parameter on the basis of satellite data (for example, Mihalcea et al., 2008), are difficult and laborious. For this very reason, measuring of supraglacial debris thickness on the Djankuat glacier took more than 1.5 months of ablation in season 2012. Measurements were taken within the range of altitudes from 2700 to 3300 m a.s.l. When debris cover thickness in the measuring point did not exceed 20 cm, the mean value of three measurements was taken there as a scale reading in the point.

It is worthwhile noting that the main advantage of the direct debris thickness measuring method lies in the fact that the researcher gains a possibility to behold the variety of the forms of debris material on the glacier snout with his own eyes. Consequently, a possibility emerges to stop thickness measuring in grid points in favor of the landscape-morphological method, when the quantity of measuring points is highly increased at difficult parts (for example, cross-profiling of debris ramparts), while diffusion of measuring points takes place on homogeneous sections (debris mantles).

Debris mapping data, obtained in 2010, was further interpolated with the use of cartographical digital packages. Methods of interpolation enable a possibility to make a model of the surface, using point by point measurements, and predict the values of variable numbers for every point of the landscape. These methods of building the digital surface are based on assumption of similarity of nearby probation points. Method of inverse factored distances was used in this particular case. As a result, the third debris thickness map was obtained over 45 years of monitoring history (Figure 2).

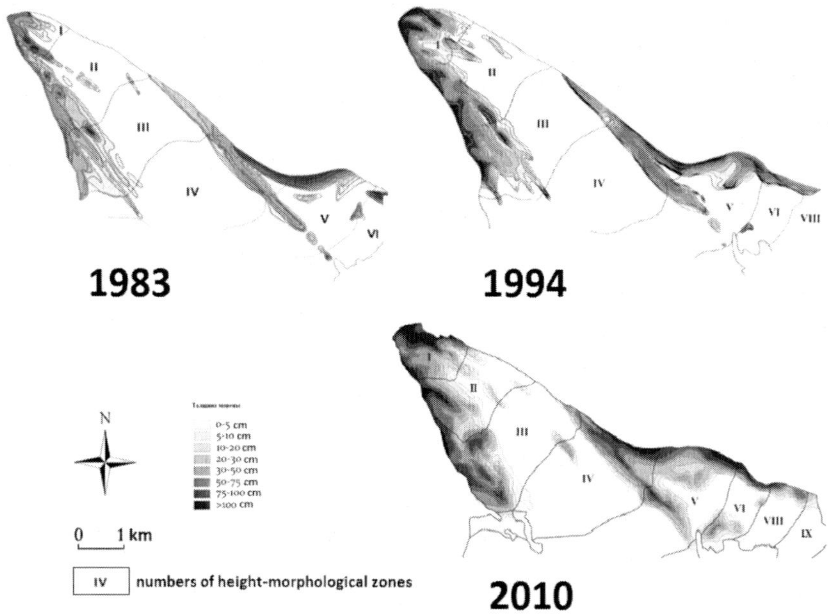

Figure 2. Debris thickness maps for 1983, 1994 and 2010.

Building of this map (obtained 16 years after the preceding one), coupled with the complex analysis of data, received during the ablation season of 2010 and previously, provided an option to trace the changes in debris cover thickness throughout the entire glacier area over 27 years (Table 1).

RESULTS

Comparison of time-transgressive maps gives an opportunity to say that the debris-covered part of the glacier is consistently expanding – 2% in 1968, 4% in 1974, 7% in 1984, 8% in 1992, 10% in 1996 and 1999, and 13% in 2010 (Figure 3).

Zone I (first) height-morphological zone of the glacier was fully armored by debris to 2010, while zone II was less than half covered. Considering the process of supraglacial debris expanding in a more detailed way, divided by height-morphological zones, it cannot go unnoticed that a naturally determined process takes place: the near-front zones of the glacier are becoming debris-covered faster and stronger than the upper ones (Figure 4).

Table 1. Mean values of debris cover thickness in different height-morphological zones

Zone number	1983 h, см	1994 h, см	1983-1994 Δh, см	2010 h, см	1983-2010 Δh, см
I	29	45	*+16*	100	*+71*
II	22	43	*+21*	51	*+32*
III	14	30	*+16*	38	*+24*
IV	34	47	*+13*	51	*+17*
V	31	32	*+1*	40	*+9*
VI	37	38	*+1*	42	*+5*

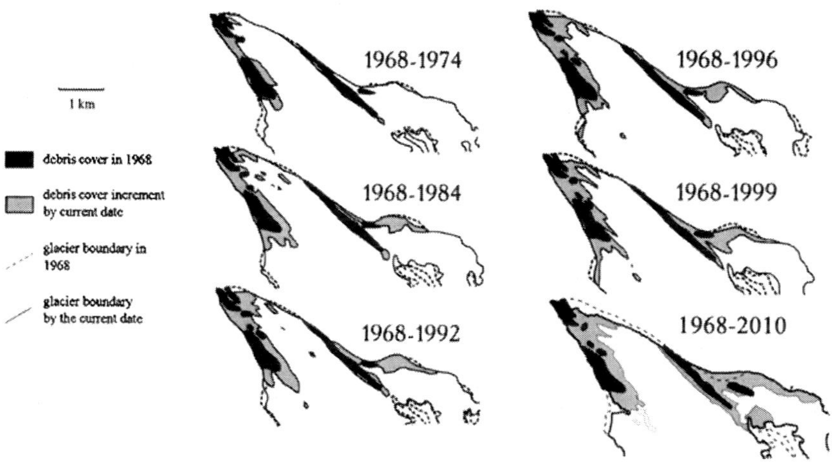

Figure 3. Debris cover expansion dynamics in 1968 – 2010.

The lowermost, zone I (2700 – 2770 m a.s.l.), was only 40% debris-covered by 1968, while nowadays it is almost fully wrapped by debris. The debris-covered area increased from 1968 to 2010 in all height-morphological zones: in zone II (2770 – 2850 m a.s.l.) – from 15% to 55%; in zone III (2850 – 2940 m a.s.l.) – from 17% to 32%; in zone IV (2940 – 3020 m a.s.l.) – from 8% to 24%; in zone V (3020 – 3120 m a.s.l.) – from 3% to 26%, in zone VI (3120 – 3200 m a.s.l.) – from 0% to 18%; in zone VII (3200 – 3300 m a.s.l.) –

from 0% to 3%. Rarely observed episodes of zonal debris-covered area compressings are explained by a full melting of the glacial core under the debris cover (i.e., the considered part is no longer a glacier, as, for instance, in zone I from 1996 to 1999) or by displacement of debris to the subjacent height-morphological zone as a consequence of ice creep.

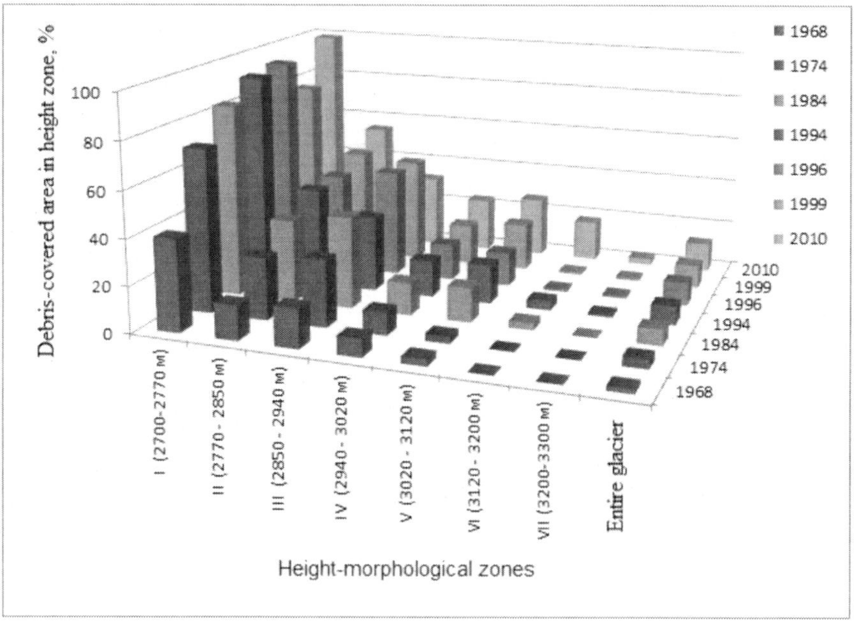

Figure 4. Dynamics of debris-covered area increment on Djankuat glacier from 1968 to 2010, divided by height-morphological zones.

Abrupt expansion of the debris-covered area in the upper zones, despite natural processes, is attributed to meltout of debris material, which entered the glacier body as a result of major rockfalls in the accumulation area in 2001-2002.

Orthogonal square of the glacier reduced from 2,99 to 2,688 sq. km for that period, i.e., the glacier lost 10% of its square for 38 years. Physical square of the glacier remained almost unchanged in 1968 and 1974 – 3,034 and 3,039 sq. km respectively – but it decreased to 2,913 sq. km in 1984 and, further, to 2,816 sq. km in 1992, to 2,767 sq. km in 1996 and to 2,740 sq. km in 1999. However, it is useful to highlight in this regard that oscillations of the total area of Djankuat glacier are affected by processes such as ice-divide migration on Djankuat firn plateau, which led to rapid reduction of the glacier square, or, conversely, compensated glacial area loss at the northern macroslope during certain time intervals (Aleynikov et al., 1999). Glacier square calculation could also be influenced by such factors as human error in defining the ice-divide border.

The glacier boundary at the snout is a lot less dynamical than at the rocky margins of firn basin. It is the consequence of a thin (first meters) ice stratum. Thereupon, predominance of negative values of mass balance on the deglaciation stage leads to the full meltdown of the ice there. This process developed most actively beginning in 1998. Consequently the quantity of rockfalls from near-ridge mountain areas increased dramatically (Popovnin and Naruse 2005), which led to thickening and aerial expansion of supraglacial debris.

Maximal thickness of debris was noticed along the axis of debris ramparts and in marginal parts of the glacier snout, where moving debris merged with deposited ones. The thinnest layer of supraglacial debris is confined to the inner periphery of the debris-covered area. If the cover is composed with material dispersed in the glacier body and cropping out to the glacier surface as a result of melting out after passing icefall and roughnesses of subglacial relief, then horizontal gradients of debris cover thickness are low, and debris ramparts are framed by relatively wide bars of thin rocky debris material. This case was observed on the left part of the Djankuat glacier. Vertical tensions are lower there, and ice movement at the approach to the snout therefore has a parallel or divergent character. Thence, there is an area in II and III height-morphological zones, where debris appears as a united lithogenous body with smoothed mesorelief features and minimal horizontal thickness gradients. The situation changes at the near-frontal part of the glacier: axial forms with increased debris thickness and huge spatial thickness variations are dominating there.

It is caused by cropping out of basal ice with a big volumetric concentration of debris material. Both studies led to obtaining a verifiable data of supraglacial debris thickness on the Djankuat glacier: it didn't exceed 183 cm in 1983 (Popovnin 2003), while in 1994 a depth of 280 cm was discovered.

The thickest debris cover was described in height-morphological zone I of the glacier: the deepest probing points (245 cm and 260 cm) were situated in close vicinity to the edge of the glacier. However, debris 35 cm and 37 cm thick were also observed in that zone. The mean thickness of debris cover in height-morphological zone I equaled 100 cm. Height-morphological zone II of the glacier is characterized by a thin debris cover and a large variety of debris relief forms: ramparts, homogenous cover and thin recently formed mantles are present there. The main volume of debris material in that zone is located near the orographically left glacier margin. The largest thickness of debris equaled 119 cm there. The thinnest debris cover was observed near the orographically right glacier margin on debris mantles, where it equaled 5-13 cm. Mean thickness of debris in height-morphological zone II equaled 51 cm. Debris cover of height-morphological zone III is represented by a homogenous surface, occasionally cut by meltwater runoff channels, where debris cover thickness does not exceed 6-13 cm. The thickest debris in this zone form layer is 101 cm. Herewith, supraglacial debris thickness quite proportionally and logically decreases with distancing from the orographically left glacier margin. The mean debris thickness in this zone is the lowest over the entire glacier and equals 38 cm. Debris cover in height-morphological zone IV is mainly located near the orographically right glacier margin in the shape of medial debris rampart, originating by confluence of two ice fluxes (Figure 5 – C and D) and furthermost relief lowering between the debris rampart and the edge of the glacier, also filled by the debris material.

Figure 5. Ice fluxes of *Djankuat* glacier.

It should also be noted that the debris cover of height-morphological zones I-III, confined to the orographically left glacier margin, is supplied both by rockfalls from the valley slopes and by the glacier flux A, significantly degrading recently (back in 1970's the uplying hanging glacier ran into that flux), which, in its turn, leads to the further intensification of rockfall activity from valley slopes, newly released from glacier mass pressure and, consequently, unstable.

Flux D, being alongside with flux C, the main debris material provider for height-morphological zones IV-VIII of the glacier, flowing down the steep slope of Uya-tau mountain, is also degrading intensely, which again induces intensification of rockfalls. Moreover, avalanches are regularly going down the slope of Uya-tau, carrying a significant amount of rock material and occasionally taking a catastrophic character, as, for example, in April 1980 and in the beginning of May 1997, when the height of avalanche fan reached 15 m. Upon conditions of winter snowiness growth and increasing winter temperatures, being observed lately, the frequency of avalanching from Uya-tau slope, including catastrophical cases, would increase accordingly.

For the above-mentioned reasons, a sharp increase of debris accumulation in height-morphological zones I-IV of the glacier should be expected in the near future.

But in the present days the minimal thickness of debris cover in height-morphological zone IV equals only 20 cm, while the maximum thickness is 79 cm and the mean zone thickness is 51 cm. Supraglacial debris of height-morphological zone V are located at the foot of Uya-tau slope (ororgraphically right glacier margin). The thickest debris layer in this zone is confined to the edge of the glacier, probably for the reason that, coupled with the rockfalls from valley slopes, the glacier flux, before going further downwards, is hitting a sidewall of rock framing, accumulating the major part of debris material there. Minimal measured thickness of debris cover in this zone equaled 9 cm (debris mantle on the steep Uya-tau footslope), maximal thickness was 73 cm (near the orographically right glacier margin). Mean zonal thickness of debris cover is 40 cm.

In height-morphological zone VI of Djankuat glacier debris are found in the form of a relatively narrow line, lying sporadically on the ice surface of sufficiently steep slopes, whereupon it is not improbable that in the nearest time a gravitational creep of debris material can occur in the underlying zones (in this case it falls into the field of exceptionally thick debris in height-morphological zone V, thereby causing even more thickening of it). However, despite the described process, the thickness of debris cover in this zone reaches a considerable 107 cm, while mean zonal thickness equals 42 cm.

A few measurements of debris cover thickness were taken in height-morphological zone VII. Debris cover appears there as a thin bar, presumably making a distinction between the flux D of Djankuat glacier and the adjacent Visyachiy glacier. Debris thickness varied from 5 to 72 cm there, while mean zonal value equaled 40 cm.

Djankuat glacier is one of the few, if not the only one, where debris cover thickness over its entire area was determined three times by direct measurements. The uniformity of used methods and succession of scientific crew provided proper comparability of the results. Matching of time-transgressive debris thickness maps results in the conclusion that the mechanism, conditioning spatial distribution of debris, didn't undergo sufficient changes for the past 27 years. The main mass of thawed-out debris still gravitates towards the axis of linear-stretched forms and glacier margins, where debris cover merges with the lateral moraine ramparts. However, lithogenic mantle mass, carried by the glacier on its surface, experienced significant increments in quantitative estimation (Table 2).

Height-morphological zone I demonstrates the most intense thickness growth, and it consistently decreases as the altitude grows. In such a manner, obligatory displacement of debris material into the lower parts of the glacier occurs. Generally, with the exception of the snout part of the glacier, debris thickness grows quite expectedly, which is graphically illustrated in Figure 6.

Moreover, judging from Table 2, we may also talk of a general increase of supraglacial debris thickness: the shift of thickness modal values towards higher ones is observed almost everywhere. This process occurs in all height-morphological zones, but it is most pronounced in height-morphological zone I, where the share of debris that is thicker than 100 cm equals 40%.

Considering the entire glacier, in 1983 and 1994 the largest zonal area (19% and 26% respectively) was occupied by the clusters of the thinnest debris material – from 0 to 10 cm.

But in 2010 the largest debris mass (23%) was already presented by a gradation of 31-50 cm. This consistent pattern also finds reflection in increasing mean glacier debris thickness from the first mapping to the last: in 1983 it equaled 28 cm, in 1994 – 39 cm, and in 2010 – a whopping 54 cm, i.e., almost twice as much as in the beginning of debris monitoring. Thus, all conducted investigations highlight an expressed thickening of debris cover on the snout of Djankuat glacier. But, what is more important, the obtained results can be extrapolated to the other glaciers of the Northern Caucasus. Representativeness of Djankuat glacier for Central Caucasus by morphological and mass-balance criteria was discovered and proved even before the start of direct glaciological monitoring about half a century ago (Boyarskiy 1978), and at the present time it is not doubted by both Russian and foreign glaciologists.

However, taking into account the intensification of debris cover investigations on this glacier in recent years, a decision was made to find out its representativeness in supraglacial debris perspective. Data of glaciers squares and squares of debris-covered parts on them, presented in the Glacier catalogue (Panov and Kravtsova 1967), was studied.

Table 2. Distribution of debris cover by thickness gradations

Height - morphological zone	Year	Debris cover thickness, cm							Mean debris thickness in zone, cm
		0-10	11-20	21-30	31-50	51-75	76-100	>100	
I	1983	32,4	14,7	10,8	22,5	15,7	2,9	1	29
	1994	26,5	10,3	12,2	16,4	12,2	6,1	16,3	45
	2010		6,7		16,7	20	16,6	40	100
II	1983	34,4	22,9	13,1	25,5	4,1			22
	1994	25,4	11	7,3	18,2	20	12,7	5,4	43
	2010	7	9,3	11,6	18,6	32,6	13,9	7	51
III	1983	53,7	18,6	19	6,6	1,7	0,4		14
	1994	28,3	22,4	16,4	13,4	9	6	4,5	30
	2010	16,3	18,6	9,3	27,9	16,3	7	4,6	38
IV	1983	27,6	13,4	14,9	24,6	4,5	9	6	34
	1994	13,5	10,8	10,8	24,3	27	8,2	5,4	47
	2010		15,7	19,8	19,8	32,3	7,3	4,1	51
V	1983	38,5	16,3	13,3	10,4	4,5	8,9	8,1	31
	1994	33,2	16,6	10,4	16,7	12,5	6,4	4,2	32
	2010	10,5	18	28,5	26	3,5	8,5	5,1	40
VI	1983	15,4	23,1	7,7	19,2	23,1	11,5		37
	1994	22,2	11,2	11,1	22,2	22,2	11,1		38
	2010			20	60	20			42
Entire glacier	1983	39	17,7	14,8	16,3	5,6	4	2,6	28
	1994	25,9	14,6	11,7	17,7	15,4	7,9	6,8	39
	2010	7,9	13,7	14,3	23,3	20,1	10,6	10,6	54

Figure 6. Mean time-transgressive thicknesses of debris cover in different height-morphological zones.

Valley glaciers are the most common among debris-covered glaciers on the Central Caucasus – 51 glaciers with total area of 180 sq. km. Djankuat also belongs to valley glaciers. Debris-covered compound valley glaciers occupy an area of 120 sq. km, and corrie-valley glaciers – 59 sq. km (Figure 7). The detailed analysis of debris-covered glaciers means squares and squares of their debris-covered parts proves the representativeness of Djankuat (Figure 8). Its entire square equals 2,7 sq. km and the debris-covered part occupies 0,3 sq. km (according to the data of 2010), while for valley and corrie-valley glaciers, forming the main mass of Central Caucasus glaciation, mean square equals 3,6 sq. km and 1.6 sq. km and the debris-covered area occupies 0,3 sq. km and 0,1 sq. km respectively.

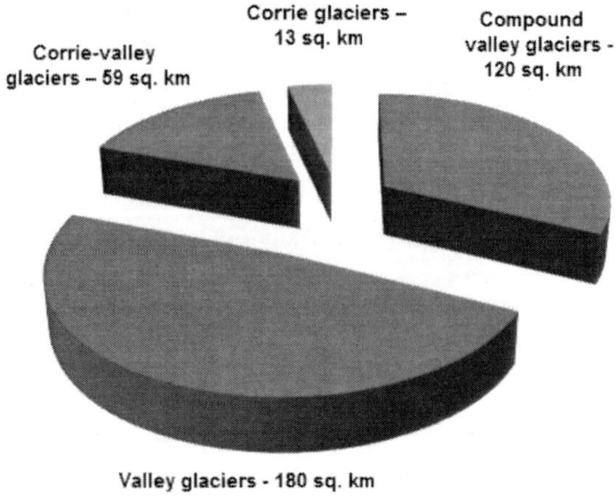

Figure 7. Total areas of different types of debris-covered glaciers on Northern Caucasus.

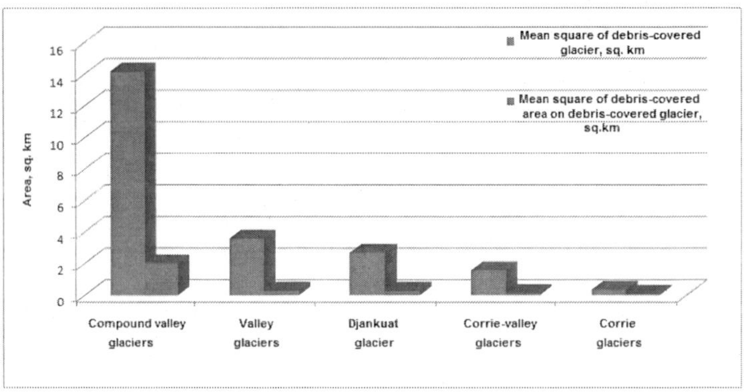

Figure 8. Mean squares of different types of debris-covered glaciers and their debris-covered areas on Central Caucasus.

Representativeness of Djankuat glacier is also confirmed by the analysis of geographical component of debris-covered glaciers distribution, averaged for river basins (Figure 9). For instance, the mean square of the debris-covered glacier in the Malka and Baksan river basins equals 3,6 sq. km while the mean debris-covered area on the glacier is 0,23 sq. km (debris-covered glaciers make 26% of the entire basin); the mean square of the debris-covered glacier in the Cherek river basin is 3,7 sq. km while the mean debris-covered area on the glacier is 3,7 sq. km while the mean debris-covered area on the glacier is

0,5 sq. km (25% of the glaciers in basin are debris-covered); in Chegem river basin the the mean square of debris-covered glacier equals 3 sq. km while the mean debris-covered area on the glacier occupies 0,3 sq. km (23% of the glaciers in basin are debris-covered), etc. On the basis of data provided in the Glacier catalogue for the Central Caucasus (Panov and Kravtsova 1967), the averaged debris-covered glacier occupies 2,4 sq. km while debris covers 0,3 sq. km of the glacier. Therefore, all the above mentioned confirms the representativeness of Djankuat glacier from a supraglacial debris perspective and provides a reason to distribute the results, obtained during the study of debris on this glacier, to the other glaciers of Central Caucasus.

The extreme poverty of direct measurements of sub-debris ablation on the other glaciers of the Caucasus keeps the glaciologist out of the reliable assessment of the universality of conclusions about the hydrological effect of superficial debris.

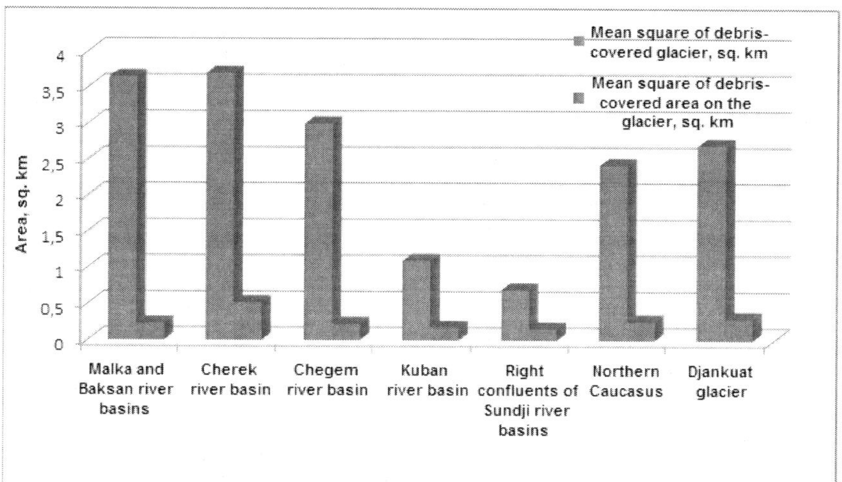

Figure 9. Mean squares of debris-covered glaciers and their debris-covered areas in different river basins on Northern Caucasus.

Predominance of debris-covered parts thicker than 5-7 cm on the Djankuat glacier clearly indicates that the screening impact of debris on ablation and runoff prevails over the amplifying effect, caused by the thin debris layer, besides background weakening of the entire glacier ablation, which varies at the rate of 20% from year to year. Similar assessment of the debris impact on glacial runoff (the only one from Caucasus) was obtained on the Zapkhito glacier in Georgia (Lambrecht et al., 2010), where the ablation weakening values turned out to be nearly identical to ones obtained on Djankuat, despite the noted differences in superficial debris cover expansion rates between the two adjacent Caucasian macroslopes (it went slower on the Georgian glacier, according to satellite data).

CONCLUSION

All conducted surveys point at an unquestionable increase of square, volume and mass of debris cover on the surface of Djankuat glacier. The map of debris thickness, obtained on the basis of direct field measurements, shows expansion of debris cover for the past 27 years both in spatial (from 0,252 sq. km in 1983 to 0,344 sq. km in 2010) and thickness (accretion up to 10 cm in the upper parts and 70 cm in the snout area) perspectives, accompanied by preservation of the main spatial principles of its distribution. The volume of supraglacial debris from 1983 to 2010 experienced 141% growth (from 70,33 to 169,59 thousands of cubic meters). Debris cover occupied 61% of glacier square in 2010. It was found that reduction of ablation due to debris occurs on 93% of debris-covered glacier surface, while debris strengthens the ablation only on the remaining 7% of the debris-covered area. Glacial runoff annually loses about 20% of its volume due to the evolution of supraglacial debris. Wide distribution of lithogenous shell at the surface certainly brings the debris evolution processes to the level of the most important factors, predetermining further evolution of the Djankuat glacier in terms of its mass balance and geometry.

In addition to the repeatedly shown glaciological representativeness of Djankuat for Central Caucasus (mainly from mass balance and morphometric perspectives), today the glaciologists have every reason to believe it is quite a representative from the standpoint of the debris cover, formed on its surface. Typicality of the debris-covered area square in the scale of the entire region glaciation is shown. Thus, the qualitative behavior of modern dynamics of the debris accumulation process and its role in the external mass-energy exchange of the glaciers can be extended to other glacial objects of the northern macroslope of the Caucasus and even (within certain limits) to the glaciers of Georgia.

REFERENCES

Aleynikov, A. A., Zolotarev, E. A., Popovnin, V. V. 1999. "Pole skorostei lednika Djankuat (Velocity field of Djankuat glacier)." *Materialy Glatsiologicheskikh Issledovaniy*, (Data of Glaciological Research) 87: 169-176 (in Russian).

Bassin, N. S., Demchenko, V. V., Sokolov L. Ya., Yanbulat, A. A. 1983. "Kolichestvenniy analiz morennogo pokrova nekotorykh lednikov Pamira (Quantitative analysis of debris cover on a few glaciers of Pamir)". *Materialy Glatsiologicheskikh Issledovaniy*, (Data of Glaciological Research) 46: 61-65 (in Russian).

Boyarskiy, I. Ya., ed. 1978. Lednik Djankuat (Djankuat glacier), Leningrad: Girdometeoizdat.. 184 pp. (in Russian).

Bozhinskiy, A. N., Krass, M. S., Popovnin, V. V. 1986. "Role of debris cover in the thermal physics of glaciers." *Journal of Glaciology*, 32, No.111: 255–266.

Lambrecht, A., Mayer, C., Hagg, W., Popovnin, V. V., Rezepkin, A. A., Lomidze, N., Svanadze, D. 2011. "A comparison of glacier melt on debris-covered glaciers in the northern and southern Caucasus." *The Cryosphere*, 5: 525–538.

Mihalcea, C., Mayer, C., Diolaiuti, G., D'Agata, C., Smiraglia, C., Lambrecht, A., Vuillermoz, E., Tartari, G. 2008. "Spatial distribution of debris thickness and melting from remote–sensing and meteorological data, at debris–covered Baltoro glacier, Pakistan." *Annals of Glaciology*, 48: 49–57.

Panov, V. D. and Kravtsova, V. I. 1967. Catalogue of Glaciers of the USSR: Vol. 8, Parts 1–4, Basin of the Kuban River. Leningrad: Girdometeoizdat. 145 pp. (in Russian).

Popovnin, V. V. and Rozova, A. V. 2002. "Influence of sub-debris thawing on ablation and runoff of the Djankuat Glacier in the Caucasus." *Nordic Hydrology*, 33, No.1: 75–94.

Popovnin, V. V., Naruse R. 2005. "A 34-year-long record of mass balance and geometric changes of the Djankuat Glacier, Caucasus." *Bulletin of Glaciological Research*, 22: 113–125.

Popovnin, V. V. 2003. "Podmorennaya ablyatsiya na yazyke lednika (Subderis ablation on the glacier snout)." *Materialy Glatsiologicheskikh Issledovaniy,* (Data of Glaciological Research) 94: 31-41 (in Russian).

Zolotarev, E. A. and Popovnin, V. V. 1993. "Gipsometriya lednika Djankuat: izmeneniya posle MGD (s 1968 po 1984). (Hypsometry of Djankuat glacier: changes after IGY (from 1968 to 1984))." *Materialy Glatsiologicheskikh Issledovaniy*, (Data of Glaciological Research) 77: 58-66 (in Russian).

INDEX

#

A

B

C

D

E

F

N

O

P

Q

R

S

T

U

V

W